Palgrave Studies in the Future of Humanity and its Successors

Series Editors
Calvin Mercer
East Carolina University
Greenville, NC, USA

Steve Fuller
Department of Sociology
University of Warwick
Coventry, UK

Humanity is at a crossroads in its history, precariously poised between mastery and extinction. The fast-developing array of human enhancement therapies and technologies (e.g., genetic engineering, information technology, regenerative medicine, robotics, and nanotechnology) are increasingly impacting our lives and our future. The most ardent advocates believe that some of these developments could permit humans to take control of their own evolution and alter human nature and the human condition in fundamental ways, perhaps to an extent that we arrive at the "posthuman", the "successor" of humanity. This series brings together research from a variety of fields to consider the economic, ethical, legal, political, psychological, religious, social, and other implications of cutting-edge science and technology. The series as a whole does not advocate any particular position on these matters. Rather, it provides a forum for experts to wrestle with the far-reaching implications of the enhancement technologies of our day. The time is ripe for forwarding this conversation among academics, public policy experts, and the general public. For more information on Palgrave Studies in the Future of Humanity and its Successors, please contact Phil Getz, Editor, Religion & Philosophy: phil.getz@palgrave-usa.com

More information about this series at
http://www.palgrave.com/gp/series/14587

Steve Donaldson • Ron Cole-Turner
Editors

Christian Perspectives on Transhumanism and the Church

Chips in the Brain, Immortality, and the World of Tomorrow

palgrave
macmillan

Editors
Steve Donaldson
Samford University
Birmingham, AL, USA

Ron Cole-Turner
Pittsburgh Theological Seminary
Pittsburgh, PA, USA

Palgrave Studies in the Future of Humanity and its Successors
ISBN 978-3-030-07990-1 ISBN 978-3-319-90323-1 (eBook)
https://doi.org/10.1007/978-3-319-90323-1

Cover design by Laura de Grasse

Printed on acid-free paper

This Palgrave Macmillan imprint is published by the registered company Springer Nature Switzerland AG
The registered company address is: Gewerbestrasse 11, 6330 Cham, Switzerland

Foreword: The Moral Dilemmas of Technology

As far back as we have evidence, humans have been using technology to extend our reach in the world. We use tools to hunt and gather food, build shelters from the elements, and communicate with others. From a theological perspective, our use of tools and technology reflects something significant about human nature. Made in the image of our Creator, we take delight in being creative, bringing about in the physical world ideas that began in our imagination. Indeed, it seems more than coincidental that Jesus was said in the Gospel of Mark (6:3) to be a *tekton*, the Greek term for an artisan from which our word technology derives. When our Savior labored in earthly form, he worked in a distinctively human occupation.

In the last two centuries, however, there has been an unprecedented growth in the number and power of new technologies. As Mark Dodgson and David Gann argue in their book on innovation, "Our great grandparents lived in a world with no light bulbs, cars, telephones, bicycles, refrigerators, or typewriters, and their lives probably had more in common with a Roman peasant than with ours. In the relatively short period of 150 years, our lives at home and work have been completely transformed by new products and services."[1] Or to use another example of the speed of change: Intel engineers calculated that if a 1971 Volkswagen Beetle had kept pace with computer chip innovation, by 2015 it would go about 300,000 miles per hour, get 2 million miles per gallon of gas, and cost four cents.[2] The

[1] Dodgson and Gann. *Innovation*, xi.
[2] Friedman. *Thank You for Being Late*, 36.

exponential expansion and acceleration of technology is hard to conceive, which makes it difficult to track the changes that our technologies are having on us. Each new generation assimilates to and grows dependent upon the new innovations—can we remember life before computers?

What happens when we develop technologies that allow us to modify ourselves, enhancing human bodies in ways that previous generations could have never imagined? Looking at the accelerating growth of technology that can alter human bodies, we may be on the cusp of another intellectual revolution—like the Copernican and Darwinian revolutions—where our conception of the universe will be radically changed. The historian Alexandre Koyré, in his classic work *From the Closed World to the Infinite Universe*, described the dramatic consequences a heliocentric conception of the universe had for understanding ourselves, moving from a finite, hierarchically ordered world to one that is cosmologically infinite.[3] In the nineteenth century, evolutionary theory challenged the view that the cosmos was perfectly created at a single moment in the past and has remained largely unchanged down to the present. Darwinism displaced the idea that living things are static or fixed in their forms, making change a fundamental feature of the natural world. Will the growth in technological control over human bodies lead us to conclude that human nature is not set in place by evolution, that we can remake ourselves in whatever image we desire? The theological questions raised by such new technologies are profound.

Many observers of technology wonder whether our species can keep up morally with our technological capabilities. In Thomas Friedman's best-selling book about the "age of accelerations," he argues that innovations in the use of machines will require moral innovation. He quotes approvingly President Obama's speech at Hiroshima in May 2016: "The scientific revolution that led to the splitting of an atom requires a moral revolution as well."[4] Friedman says we will need "to reimagine how we scale sustainable values to everyone we possibly can when the power of one and the power of machines become so amplified that human beings become almost godlike."[5] From a Christian perspective, this call to moral innovation will be met with justifiable skepticism. Have humans ever dis-

[3] Koyré. *From the Closed World to the Infinite Universe.*
[4] Friedman, *Thank You for Being Late*, 349.
[5] Ibid., 199.

played the consistent ability to overcome selfishness and greed through moral effort? Judging based on human history, our sinful natures are not so easily dislodged.

This then is the dilemma the church faces in the modern age: at its best, modern technologies provide new opportunities for human flourishing by overcoming frailties and alleviating pain and suffering. At their worst, modern technologies provide opportunities for humans to gain selfish advantage over our neighbors and to push foolishly beyond the limits set by our biological natures. Should Christians promote the responsible alteration of human bodies because we are utilizing powers made possible by our God-given capacity to reason? Or should we resist the temptation to make choices for what should be reserved for God alone?

I do not know the answer to these questions. I do not even know if there is a single answer to give, because each technological innovation raises unique questions. But I know that the church needs to think deeply on such matters so that we may respond with wisdom and discernment. I am thankful to the authors and editors of this book for putting the issue before us, so that we may consider how to better practice Christian discipleship in a technological age.

Birmingham, AL, USA Josh Reeves

References

Dodgson, Mark, and David Gann. 2010. *Innovation: A Very Short Introduction*. Oxford: Oxford University Press.

Friedman, Thomas. 2016. *Thank You for Being Late: An Optimist's Guide to Thriving in the Age of Accelerations*. New York: Farrar, Straus, and Giroux.

Koyré, Alexandre. 1957. *From the Closed World to the Infinite Universe*. Baltimore: Johns Hopkins University Press.

Contents

Notes on Contributors

Christina Bieber Lake is the Clyde S. Kilby professor of English at Wheaton College where she teaches classes in contemporary American literature and literary theory. Lake has authored several essays on Flannery O'Connor as well as a monograph entitled *The Incarnational Art of Flannery O'Connor* (Mercer UP, 2005). Her most recent book *Prophets of the Posthuman: American Fiction, Biotechnology, and the Ethics of Personhood* (University of Notre Dame Press, 2013) was awarded Indiana Wesleyan University's Aldersgate prize for integrative scholarship. She is at work on a project that examines the theological moorings of contemporary American literature.

Jeffrey P. Bishop is Professor of Philosophy and Professor of Theological Studies at Saint Louis University, where he also holds the Tenet Endowed Chair in Health Care Ethics. His work examines the social, political, historical, and philosophical conditions and justifications for contemporary medical and scientific practices and theories. In addition to numerous journal articles, he has authored the book *The Anticipatory Corpse: Medicine, Power, and the Care of the Dying*. He serves on the editorial boards of the *Journal of Medicine and Philosophy*, *Christian Bioethics*, and the *Journal of the Society of Christian Ethics*.

Ron Cole-Turner teaches theology and ethics at Pittsburgh Theological Seminary, where he holds the H. Parker Sharp Professorship. He is a founding member of the International Society for Science and Religion, which he serves currently as a vice president. He is the author and editor of several books on science, technology, and Christian theology, from *The*

New Genesis: Theology and the Genetic Revolution (Westminster John Knox, 1993) to *The End of Adam and Eve: Theology and the Science of Human Origins* (TheologyPlus, 2016). He is at work on a book on human enhancement technology and its impact on theology.

Michael Dickson is Professor of Philosophy in the College of Arts and Sciences at the University of South Carolina, where he has been since 2004. Prior to moving to South Carolina, he was Ruth N. Halls Professor of History and Philosophy of Science at Indiana University. He works primarily in philosophy of science, especially the foundations of quantum theory, but has also done work in general philosophy of science and the relationship between science and metaphysics. More recently, his work has been in game theory, and specifically in evolutionary game theoretic models of linguistic and epistemological development.

Steve Donaldson is director of the computer science program and co-director of the computational biology program at Samford University. He also teaches in the University's Fellows and Science and Religion curricula and is a co-founder of the Samford University Center for Science and Religion. His professional interests include cognition, models of intelligence, autonomous systems, self-organization and emergence, and the interface of science and religion. He has authored *Dimensions of Faith: Understanding Faith through the Lens of Science and Religion* (Cascade, 2015) and co-authored (with theologian Josh Reeves) *A Little Book for New Scientists* (IVP Academic, 2016).

George Michael Glawson is a Ph.D. student in philosophy and instructor at the University of South Carolina. His academic work lies at the intersection of ethics, technology studies, and phenomenology, where he focuses on the complex ways our values, our technologies, and the intricacies of our first-person experiences shape one another. He also serves as a consultant, helping company leaders think profitably and kindly about issues of ethics, engineering, and employee well-being. His spare time is spent working on acrylic sculpture, expanding his collection of old and strange books, and playing drums for artists and studios in Columbia, Birmingham, and Atlanta.

Fred Glennon is professor and chair of the Department of Religious Studies at Le Moyne College. His research and teaching focus on religious ethics, social justice, and public policy. His essays have been published in the *Annual of the Society of Christian Ethics*, the *Journal of Catholic Social*

Thought, and the *Journal of Religious Ethics*. He is coauthor *of Introduction to the Study of Religion* (Orbis Books, 2012), now in its second edition. He also publishes in the area of the scholarship of teaching and learning, with a number of publications in *Teaching Theology and Religion*.

Boaz Goss is a PhD student in the dual Heath Care Ethics/Theology program at Saint Louis University. His mentor is Jeffrey P. Bishop. Boaz's primary background is in Christian philosophy and he is most influenced by Plato, St. Aquinas, and Nietzsche. His dissertation project seeks to harmonize a traditional Christian understanding of the soul with modern neuroscientific studies on free will. He hopes to demonstrate that—contrary to popular belief—modern neuroscience has not "disproven" free will, but actually coheres nicely with traditional Christian thought. Boaz is also interested in the philosophy of music and philosophy of dance.

Allison Hepola earned her PhD from Rutgers, where she focused on metaphysical issues surrounding literary interpretation.

Ysabel Johnston is a PhD student in health-care ethics and theology at Saint Louis University, and received her Bachelor's in Philosophy at Azusa Pacific University. She uses resources from phenomenology, feminist philosophy, and mimetic theory to address bioethics issues relating to religion, technology, and bodily self-perception. Her dissertation will analyze the relationship between theological accounts of natural evil and bodily identity in the context of health care. She recently received the Young Scholars Award at the 2017 Societas Ethica conference. When she's not writing, Ysabel enjoys singing at open mics and watching local theater productions.

Steven J. Kraftchick is Professor of the Practice of New Testament Interpretation at the Candler School of Theology and the Graduate Division of Religion at Emory University. His teaching, research, and writing focus on the history of New Testament interpretation, New Testament theology, and the study of Paul's letters. His current research interests include contemporary interpretations of Paul's epistles and the intersection of theology and technology, especially as it relates to conceptions of the "techno-human" person and the philosophy of technology. He is currently working on Pauline conceptions of the embodied human in relationship to both individual and communal selves.

Josh Reeves is assistant professor of Science and Religion at Samford University. Before coming to Samford, he completed a two-year postdoctoral fellowship in the Heyendaal Program on Theology and Science at Radboud University in the Netherlands. He is author of *Against Methodology in Science and Religion: Recent debates on rationality and theology* and co-author of *A Little Book for New Scientists: Why and How to Study Science*.

Jeanine Thweatt holds a doctorate in Theology and Science from Princeton Theological Seminary. She is the author of *Cyborg Selves: A Theological Anthropology of the Posthuman*. Her research interests focus on the overlap between theology, philosophy, and the sciences, and particularly on the definition of the human from these various perspectives. She is currently Visiting Lecturer in Humanities at Flagler College in St. Augustine, Florida.

Carol Ann Vaughn Cross is assistant professor at Samford University where she teaches social history. She earned her doctorate in history at Auburn University in the fields of religion, technology and culture, and women's history. She has other publications appearing in *Social Sciences & Missions*, *North American Foreign Missions, 1810–1914: Theology, Theory, and Policy*, and EthicsDaily.com. She has served on the National Coordinating Council of Cooperative Baptist Fellowship and as an ordained elder in the Presbyterian Church USA. She previously directed the Christian Women's Leadership Center and taught history at Judson College for Women.

CHAPTER 1

Introduction: Why the Church Should Pay Attention to Transhumanism

Ron Cole-Turner

Why Christians Should Care About Transhumanism

With all of the problems in the world today, why would anyone suggest that Christian churches should care about transhumanism? It is a small movement, barely a blip on the cultural landscape. As an organization, transhumanists have at most a few thousand followers, mostly techies and philosophers of technology. They have an academic journal, a website, and an institutional toehold at least for now at Oxford University.[1] They grab

[1] The main transhumanist journal is the *Journal of Evolution and Technology*, now in its 27th year and published electronically by the Institute for Ethics & Emerging Technologies (IEET) on the Web at http://ieet.org/. The site includes many resources and a frequently-updated blog. The Future of Humanity Institute, located at the University of Oxford, hosts a research team and provides resources on transhumanism and enhancement technology at its website at https://www.fhi.ox.ac.uk/). IEET and the Future of Humanity Institute use Twitter, with about 8000 followers each.

R. Cole-Turner (✉)
Pittsburgh Theological Seminary, Pittsburgh, PA, USA
e-mail: coleturn@pts.edu

S. Donaldson, R. Cole-Turner (eds.), *Christian Perspectives on Transhumanism and the Church*, Palgrave Studies in the Future of Humanity and its Successors, https://doi.org/10.1007/978-3-319-90323-1_1

the headlines from time to time with promises that seem to be too good (or too scary) to be true. Even so, most people have never heard of transhumanism. Or if they have heard, they are not particularly interested in hearing more. So why should churches pay any attention?

Transhumanism deserves attention because of what it points to. The focus of this movement is the future of technology, especially its power to enhance human beings and to create artificial intelligence or AI. The technological transformation of humanity has already begun, and the changes we can see today are already making their mark on the social, psychological, and cultural fabric of our time. At the same time, advances in the field of computing are leading to powerful new forms of AI. Some experts believe that we will soon be creating machines more intelligent than any humans. To their credit, transhumanists urge us to pay attention to the humanity-transforming and humanity-displacing powers of technology. That alone makes them good to have around.

Some enhancement technologies, of course, are already so obvious that we do not need transhumanists to point it out. The best known is the use of various technologies to enhance athletic performance. New training methods, new equipment, widespread use of supplements, and of course performance enhancing drugs are all part of sports today, from Olympic arenas to suburban peewee football fields. Some drugs are legal to use when obtained with a prescription associated with a properly diagnosed medical condition. Some substances that do not require a prescription might be banned by sports regulatory authorities. Restrictions notwithstanding, many athletes turn to performance enhancing substances, sometimes justifying their decision because they are competing against other athletes who use similar substances.

It may be no coincidence that the public is mostly aware of advances in supercomputers because of well-publicized sports-like competitions that pit intelligent machines against human champions. Some of the best-known computer-vs-human competitions include chess, the traditional Chinese game called "Go," and the television game "Jeopardy." The machines are winning, but of course we need to bear in mind that a computer's ability to beat the grand champion in the defined space of a game is not the same as exhibiting overall intelligence superior to humans. These competitions suggest, however, that AI research may be closing in on that goal.

Some of these developments are visible to the wider public. Many more advances in technology, however, are occurring daily in university and corporate research centers around the world. What the transhumanists help us see is that enhancement in sports and computer success in games like Jeopardy are just small parts of what is going on. Human enhancement technology pervades nearly every dimension and arena of our lives today, from the military and the workplace to academia. When we take the time to look closely, we can see people using technology to boost their performance in almost every field of human activity. Transhumanists see the broad connections here. They see human enhancement through technology as a system-wide trend, affecting nearly every human capacity and every dimension of human performance. They also wonder about the future of AI and whether some form of human consciousness could live indefinitely inside the intelligent machines that we create.

The fact that transhumanists like what they see here is no reason to dismiss the reality of the trend-line to which they are calling our attention. Looking not just at sports but in every direction, what we see is that the development of the technologies of human enhancement is already moving forward, gaining speed as it goes along. The trend line of technological advance is real. Does anyone really think the trend is about to stall or go in reverse? All the signs indicate that advances in technology are more globally distributed, more heavily funded, and more quickly achieved than ever before. The effects will be utterly transformative, changing forever what it means to be human.

What technologies will contribute to this transformation? Today it is mainly pharmaceutical products, developed of course to treat disease but used by many to enhance human performance. For example, more than a fourth of the students in universities say they have used prescription drugs such as Adderall, not because they have their own prescription or are trying to treat a learning problem, but because they want to boost their academic focus and their ability to concentrate for long periods of intense study. Many adults today use supplements such as resveratrol to tweak their metabolism, hoping to delay the aging process and increase longevity.

Do these drugs and supplements actually enhance human cognition or longevity? The evidence today is inconclusive. But in all probability, new and more effective drugs will be developed and used. Furthermore, the enhancement technologies of the future are likely to include many things beyond drugs. Transhumanists like to call our attention to things like chip implants, gene editing, and nanotechnology, all of which they say are likely to play a role in helping us transcend our biological limits. When these technologies are more fully developed, transhumanists insist, these new powers will make today's attempts at human enhancement seem small.

Today's small and partial enhancements point to what is coming. When technology advances step by step, it is too easy for us to get used to each step as it comes along and to overlook the net effect. We find ourselves very much like the proverbial frog in a pot of water warming slowly on the stove. We need to pay attention to the technological temperature of our times. Already today, technology is changing us. The greatest danger here is perhaps not the technology itself but our lack of awareness of what it is doing to us. Transhumanists are very aware of what is going on, and talking with them can heighten our awareness of how we are already being changed, step by step.

And beyond that, transhumanists offer us a preview of one possible version of the culture of the future. They really like technology. But more than that, they like what it can do to them. Others may worry that technology is dehumanizing us, but transhumanists embrace technology precisely because it super-humanizes us, at least as they see it. They call themselves transhumanists because they want to use technology to make themselves more than human. They predict that, over time, many of us will fall in love with what enhancement technology can do for us. When that happens, we will convert to their point of view. They think that when technology becomes more pervasive, more powerful, and more widely used, more people will be transhumanists in all but name.

Transhumanists have a point here, and we can see evidence for what they are saying by watching how much people love their computers and smartphones. Many people today wonder how they ever got along without these things. And when they think about it, they wonder how they could possibly manage anymore without modern medicine or transportation or other technology-based conveniences. Transhumanists predict that in the future, many of us will feel the same way about an ever-widening array of technologies, including those that act directly on the human body and brain. Their prophecy is that when we are enhanced, the new level will seem normal and we will not even be able to imagine going back.

Most of all, transhumanists are very helpful in mapping out scenarios of the human future. They work hard to imagine as concretely and specifically as possible how various technologies might develop, how they might affect individuals, and what the positive and negative social and economic impacts might be. We may think their predictions are wrong. The value of transhumanists for the church, however, does not rest on whether the future will match their scenarios. Their value lies in the way they stretch our imaginations, challenge us to develop our own ideas about the future, and most of all show us how seriously we ought to take human enhancement technology in all our assessments of what is coming. The futuristic scenarios of the transhumanists are an open invitation for the church to think in its own way about the culture of the future.

What is it, after all, to "discern the signs of the time"? We are admonished not just to pray but to "watch and pray." It goes without saying that watchful Christians today have to be alert to many trend lines. The economic, social, and political dramas of our times, played out in small towns or on the global stage, are all at work reshaping the world in which we live. But among all the varied changes at work remaking our world, nothing is so determinative in the long run as technology broadly considered. Its pace is relentless, even accelerating. Its reach is global, unconstrained by borders or traditions. The arena of its action extends from communications to agriculture to energy and far beyond. It affects all other dimensions of human social and economic change. It changes everything around us. And today, it is beginning to change everything *inside us.*

Basic biological functions, brain activity, emotions, cognitive capacities, and moral predispositions are all susceptible to the inward reach of new and sophisticated technologies. The church today cannot ignore this. Technology is changing the very notion of what it means to be human. We may not agree with the transhumanists, but they can be very useful to have around because they show us where to look and what to watch. Their prayers may be different from ours, trusting as they do in technology for something like salvation. But when it comes to the "watch" part of "watch and pray," they are right on track.

As a movement, transhumanism itself may not be all that important. What is important is what it points to, which is the transformation of humanity through the use of the technologies of human enhancement and the concomitant rise of AI. Quite simply, it is not an option for churches and theologians to ignore these trends. One way to focus our attention is to engage transhumanism. Transhumanists enthusiastically proclaim a gospel of human enhancement and the coming blessings of AI. In this chapter, therefore, we will use "transhumanism" as a kind of shorthand for *an exuberant embrace of AI and of human transformation through technology in the hope of going beyond the biological or physical limits of humanity*. By engaging transhumanism, Christians force themselves to rethink the meaning of the gospel we proclaim and the shape it is taking in our technologically-saturated times.

It is also worth noting here at the outset that while technology is global and transhumanism is a world-wide movement, this book draws upon and reflects the debate currently underway in the United States. Readers in other countries will of course be aware that the US is at once one of the most technological and most religious of all nations. For that reason, how religious people respond to technology here will not be the same as in other countries where the practice of religion may be less widespread and publicly visible. This book is also limited by the fact that all the contributors are Christian. It is not a book on "Religion and Transhumanism." Our goal here is limited to trying to reflect the conversation underway among American Christians. Given the sheer numerical prominence of Christians in the US and their social and economic impact, however, their response to technologies of human enhancement has consequences far beyond our national boundaries.

Christians Responding

Perhaps even more interesting is the fact that Christians in the US differ dramatically among themselves in how they regard transhumanism. That conflict is reflected in the chapters that follow. It is also evident in recent public polling data. If we ask how Christians respond to transhumanism today, the answer is complex. Some object to everything it stands for. Others embrace it. Most are ambivalent, liking technology's benefits but worried about where it might take us. All this is reflected in a 2016 survey conducted by the Pew Research Center on attitudes about human enhancement technology. According to the Pew report, "Cutting-edge biomedical technologies that could push the boundaries of human abilities may soon be available, making people's minds sharper and their bodies stronger and healthier than ever before." According to the survey data, however, most Americans "greet the possibility of these breakthroughs with more wariness and worry than enthusiasm and hope."[2]

[2] Funk, et al., "U.S. Public Wary of Biomedical Technologies to 'Enhance' Human Abilities."

Among other things, the study found that Christians in America are concerned about enhancement technology at roughly the same level as the general public. Differences in attitudes about technology appear to depend more on how highly involved one is in religion than on the religion one follows. When asked about the use of gene editing for the next generation, 64% of Americans who claimed to be "highly religious" opposed this technology compared to only 28% of those who said their religious commitment was "low." Those who identify as "white evangelical" were 63% in opposition compare to 20% or lower among those who identified themselves as atheists or agnostics.[3]

These differences are significant, but it is important not to read too much into these findings. They show that a significant majority of Americans, including most Christians, are worried about these technologies. It is plausible to think that these worries will translate directly into a negative assessment of the transhumanism movement. The findings also show, however, that not all Christians are opposed to enhancement technologies such as human germline modification. For example, among white evangelicals (the block of Christians most opposed), 63% stood in opposition. But 32% of white evangelicals found certain enhancement technologies acceptable at least under some circumstances. When asked by pollsters, they agreed that human germline gene editing "is no different than other ways we try to better ourselves."

A few Christians, in fact, see themselves as transhumanists, going so far as to call themselves "Christian Transhumanists." Some have created an organization, the "Christian Transhumanist Association," which they launched in 2014. They invite people to join them by endorsing a brief "Affirmation." Central to the Affirmation is the belief that God is at work in the world in "the transformation and renewal of creation including humanity," and that Christians are invited to participate in that work. At a personal level, this means seeking "growth and progress along every dimension of our humanity: spiritual, physical, emotional, mental." Science and technology are affirmed "as tangible expressions of our God-given impulse to explore and discover." In particular, the Affirmation endorses "the intentional use of technology, coupled with following Christ." The result, they believe, is that we will "become more human across the scope of what it means to be creatures

[3] Ibid.

in the image of God."[4] In 2015 the Association created a Facebook group that now has over 800 members.[5]

Whether this group thrives or disappears in the years ahead is hard to predict. If it does thrive, will it be seen by churches and other Christian organizations as mainstream or marginal? No matter which way these institutional realities play out over time, two things are clear at least for now. First, this is a tiny movement, so small it hardly merits attention at all. Second, its size notwithstanding, this organization has staked out a place in the cultural landscape. Anyone doing a web search for "Christian transhumanism" is sure to find them and to discover that at least one organized group of Christians exists for the purpose of advocating a Christian version of transhumanism. Simply by existing, this group creates at least a hint of legitimacy to the idea that Christians can embrace transhumanism.

From the group's "Affirmation," it is clear that its leaders are approaching the question of technology with theological seriousness. They affirm technology, not because it is trendy or provides them with high-tech jobs, but because they believe that God the Creator works through all things, including technology, to achieve God's own ends for creation. They affirm human enhancement, not because they want to escape the limits of biology in order to pass beyond this mortal flesh into some sort of superhuman future, but because they believe that it is God's will for human beings to advance or grow so that we might reach a form of human maturity beyond the present. They embrace technology as one means of this advance, not because they reject the grace of God in Jesus Christ, but because they believe this grace can and does work through technology as well as other means.

If a few Christians embrace enhancement technology and transhumanism, there seems to be no doubt that a majority of Christians reject the idea that there can be such a thing as "Christian Transhumanism." The most obvious difference between these groups is their attitude toward technology. Christian Transhumanists are optimistic or at least positive in how they view the effects of enhancement technology. This optimism is definitely not shared by other Christians, or indeed by the majority of Americans, who say they worry about the very idea that we will try to change our human nature through technology.[6]

[4] https://www.christiantranshumanism.org/affirmation

[5] As of January, 2018.

[6] Funk et al., "U.S. Public Wary of Biomedical Technologies to 'Enhance' Human Abilities."

Why do Christians hold opposing views on transhumanism and on technologies of human enhancement? Why do the contributors to this volume argue, at least in some cases, for precisely opposite points of view? Differing attitudes toward technology may be the most obvious difference between them. Another explanation may be that our authors hold differing views on what transhumanism means or where it originates. For some, it means little more than enthusiastic support for technology, including technologies of human enhancement. For them, transhumanism's origins go back to people like Francis Bacon, whose views are rooted in a Christian vision and constrained by it.[7] Other contributors to this book tend to see transhumanism as excessively libertarian in nature, rooted perhaps in Friedrich Nietzsche's "will-to-power," and almost by definition the antithesis of Christianity. Both versions of transhumanism's origins are held by today's transhumanists. Neither view can claim to be the "right" interpretation, but the choice between them makes a world of difference, especially for Christians.

The most *important* difference in perspective between our authors, however, is theological. How should we see God as present and active in the creation now? Christians who embrace transhumanism tend to believe that God is not entirely done with the work of creation but is actively creating even now. To this they add the even more controversial idea that one way God creates is through technology. God uses the whole creation as a means to sustain, renew, and advance the creation. For the Christian transhumanist, this suggests that if God uses everything in creation for God's purposes, then God must be able to use technology to bring creation to its consummation. They are careful here, of course. They do not want to suggest for example that technologically advanced weapons of mass destruction are used to create. But many other forms of technology are used by God. For example, medicine is used by God to bring healing. Advanced agricultural techniques are used to feed the world. Human enhancement technology, by analogy, can be seen as a way in which God is completing the work of creating humanity in its fullest and final form.

[7] For more on transhumanism's roots in Bacon and on Bacon's theological assumptions, see Joseph Wolyniak, "The Relief of Man's Estate: Transhumanism, the Baconian Project, and the Theological Impetus for Material Salvation," in Mercer and Trothen, *Religion and Transhumanism: The Unknown Future of Human Enhancement* (Santa Barbara: Praeger, 2015), 53–69.

Christians who oppose transhumanism are likely to reject one or the other of the theological beliefs embraced by the Christian transhumanists. Perhaps those opposed see creation as complete and not as something that is still being completed. God's activity in the creation today is mainly a matter of keeping the creation in existence and preventing its fall into greater disorder or sin. More likely they have their doubts about whether God uses technology to create. God may still be creating, for instance in turning this creation into the new creation. But God does not need our help to do this. They do not reject technology, and they do not rule out altogether the possibility that God may use it for God's own gracious purposes. They agree, for example, that God uses medicine for healing. But it goes too far, they think, to suggest that God uses our technology to complete or enhance the creation. Technology cannot go to the root of the human problem, which is spiritual in nature and has to do with the human propensity to rebel against God. And because it cannot undo our proclivity to warp ourselves and nearly everything around us, technology is always bedeviled by what it cannot fix. It may be very helpful in dealing with specific problems. But in the end, many Christians wonder whether technology expands rather than reverses the central problem of human sinful behavior. For reasons like these, they reject the idea that there can possibly be such a thing as "Christian Transhumanism."

This book offers criticism of both views. Have Christian Transhumanists fallen too much under the spell of the technological thinking of our age, which sees every problem as nothing more than a technological challenge? Perhaps their view tends to fit a bit too neatly with a progressivist, techno-optimistic view of the world. Where is sin in this theology, or the catastrophes of history made possible in part by technology? Above all, where is the cross of Jesus Christ? Clearly, not all technological advances and applications are being used by God to advance the work of creation. But how does one decide what is in and what is not?

On the other hand, have those who have completely rejected transhumanism fallen under a different spell, one that makes nature (or the original creation) normative in all matters of value, almost to the point of ruling out the very possibility that even God can make creation a new creation? Perhaps their view tends to fit too comfortably with the anti-technology or secular "bio-conservative" voices of our time and their techno-pessimistic view of the world. Where, then, is the grace of God at work making this creation a new creation? Christian theology cannot tolerate any hint of a partition between God and creation, or more precisely between a part of creation that is God's and a part that is not. It is not possible theologically that there is anything in creation (technology) that is not really part of God's creation and not under the rule of Jesus Christ, such that it cannot be used for God's ultimate purpose in the transformation of the creation. Our hope is in the One who *makes all things new*.

And so the arguments go. Some of the contributors to this book criticize one approach or the other, rejecting the very notion of Christian Transhumanism or its polar opposite. Many of the contributors, however, are asking whether there is a third option. If Christians neither embrace nor condemn transhumanism, can we engage it? Can we learn from it even as we criticize it? Most of all, can we use it as an occasion to reflect on the most profoundly significant cultural changes of our time? Can we let it be for us a generative provocation that challenges us to do our own best thinking about how God is active and present in this world?

The challenge taken up in the chapters that follow is twofold. First, how do we bear witness today to the God of Christian faith, the God who is at work even now in all dimensions of creation? Technology is part of the vast created order. Nothing exists outside the realm of creation or the sphere of God's sovereignty. Technology does not lie outside the scope of creation, doing whatever it wants in some alternate universe that is alien to God's own creation. We have no choice but to say that God is creatively and redemptively active in and through human technology to fulfill God's own plans for the transformation and consummation of all things in Jesus Christ.

Second, how do we live as Christians, constantly surrounded as we are with the seductions of techno-culture? It is all too easy today to succumb to the idea that all our problems can be fixed by technology. Some of them can, but deliverance from the deepest and most deadly human problems lies beyond the capacities of technology, today's or tomorrow's. Some of our problems might be addressed significantly by human enhancement. But in the end it might be our problems and not our lives that are enhanced. How can Christians find the grace to live wisely?

These are not easy questions. No contributor to this book claims to have the right or final answer. But together, by debating and exploring these questions, our authors combine their thoughts to create a kind of shared invitation to the church: Consciously and deliberately, join us in taking up the task of becoming a courageous church, relevant to our times, and able in the face of the most advanced technology to bear witness to the creative and redemptive presence of our God.

Wrestling with the Future

In the first section of this book, we discover that biblical texts focus more often than we think on the relationship between religious practice and technology. Turning to texts such as Genesis 3–6, Exodus 25–40, and Revelation 21–22, Steven Kraftchick shows how these passages pay close attention to technological details and to their theological significance. Allison Hepola shifts our focus to early Christian theologians and to their theological ruminations about human-like creatures that were not quite human. While these ancient "Dragons and Dog-Headed Saints," as they were once called, are clearly not the same as post-human forms of humanity or artificially intelligent machines, the responses of early theologians such as Augustine provides a way for us to think about post-humans or other intelligent creatures that may come along in the future.

In the next chapter, Carol Ann Vaughn Cross calls our attention to the popular nineteenth-century clergyman, Henry Drummond. In the decades of the 1880s and 1890s, amidst all the technological and scientific changes of his time, Drummond was able to offer a bold and affirmative interpretation of evolution and technological progress while also insisting on the need for an active and disciplined practice of the Christian life. The final chapter in section one is by Boaz Goss, who points to the way in which Christianity has allowed itself over the centuries to be backed into a kind of corner when it comes to engaging science and technology. Any effort by Christians today to engage transhumanism, he suggests, is rigged from the outset. Christians today need to reengage the natural world not just as creation but with an openness to the sacramental way in which it is full of God's presence.

The book's second section begins with a careful analysis by Michael Dickson of traditional Christian concepts such as the human soul and the belief that humans are created in the image of God. Dickson considers a wide range of arguments for and against transhumanism and its compatibility with Christianity, including the idea that Christians might agree that human improvements seem like a good idea until they reflect on our moral and spiritual capacity to tell real improvements from false ones. The next chapter is by Jeffrey Bishop, who is highly critical of transhumanism. He calls our attention to the metaphysical presuppositions that underlie not just transhumanism but medicine itself. Is the human body really just morally neutral and subject to the manipulations of technology's power? Bishop counters by arguing that the Christian doctrine of the resurrection establishes the counter-belief that bodies matter.

Christina Bieber Lake is also highly critical of transhumanism, but she approaches the subject through her discipline of literary criticism. Transhumanists, she insists, simply cannot tell a good story. She examines a recent transhumanist novel by Zoltan Istvan, entitled *The Transhumanist Wager*, arguing that its defects as a novel are compounded by the inability of transhumanists to tell any kind of coherent and credible story about authentic human flourishing.

The chapter by Steve Donaldson concludes this section. Donaldson begins by recognizing that cognitive enhancement is possible and is already being achieved by various means. But it is transhumanist hype, he argues, to think that an engineered form of intelligence can be all-knowing or godlike. There is no escaping the need for faith, which is always limited in the sense that it cannot escape being probabilistic. Christians need not fear enhanced intelligence, for no matter how far it advances, it does not diminish God or the creature's need for faith.

The final section opens with a chapter by George Michael Glawson. He invites us to drop back from transhumanism as a matter for debate and look at it instead as an indicator of the future. He believes that much of what transhumanists hope for is very likely to occur. If so, how does one live as a Christian in such a world, surrounded by enhanced humans and intelligent machines? If there are benefits, how do we maximize them and share them fairly? How do we mitigate harms? We are called, he suggests, to shape the future and not merely discuss it.

Jeanine Thweatt's chapter immerses us in the world of superheroes and their rivals, super villains, inviting us not only to understand the fictional world of today's youth and its video-based gaming culture but to explore how enhanced creatures become a force for good or evil in the world. How can Christians (and Christian parents!) make sense of this world? As Thweatt suggests, these science fiction narratives may be our best window into the culture that awaits the technologies of tomorrow.

The chapter by Fred Glennon argues that Christians cannot escape their connection with transhumanism. It is our shadow, Glennon suggests. It reveals the dark side of human aspirations. But more than that, transhumanism is Christianity's "constant companion" because both are rooted in profound hopes for the transformational future. Our hopes may be different, but the openness to transformation means that we are tied together in thinking about how to contribute all we can to making the future just and safe for all. Finally, Ysabel Johnston calls our attention to the ground of our yearnings for the future, asking how our deepest desires are shaped by social and cultural forces. It is naïve, she argues, to think that what we desire is simply good. Our own desires are not neutral, and in fact they are not even our own. The first step is to recognize this, and the second is to seek a reshaping of desires. For Johnston, and indeed for all Christians, this reshaping comes through the grace that transforms us to be like Christ, whose desires must become our own.

The book ends with brief epilogues that asks what we have learned and how it will help Christians and congregations in the future. Speaking in a very personal way, the two editors of this book sum up their hopes for the future of the church as it finds itself almost overtaken, surrounded, and even absorbed into the technological transformations of our time. What, indeed, does it mean today to "watch and pray"?

A Concluding Prologue

This book is not an end in itself. All of us who contributed to these pages are looking beyond the mere publication of another volume. Our hope is that this book launches a conversation in churches and among Christian leaders. We are not trying to win anyone to a particular point of view, least of all about transhumanism. Our purpose is to direct the church's attention, which can be drawn in so many different ways these days, to the trend toward human enhancement technology and to the profound changes that will follow in terms of what it means to be human. Our shared hope as contributors is that Christians everywhere will become increasingly aware of this trend and its impact on ourselves, our families, our communities, our congregations, and our world.

As important as that awareness might be, our hope goes even further. Yes, it is important that the church be awake to what is going on around us and in us. But far more important is that through the faithfulness of the church, God's purposes for the future of humanity will be more fully realized. It is not enough for Christians merely to be aware of what is going on. It is not enough to be conscious of how technology is transforming the rhythms of our lives or the patterns of our families. The church is called by God to be the place where God's future is taking shape. In other words, to discern, watch, pray, … and, with courage and faithfulness, act!

Bibliography

Funk, Cary, Brian Kennedy, and Elizabeth Podrebarac Sciupac. 2016. U.S. Public Wary of Biomedical Technologies to 'Enhance' Human Abilities. *Pew Research Center*, July 26. http://www.pewinternet.org/2016/07/26/u-s-public-wary-of-biomedical-technologies-to-enhance-human-abilities/

Mercer, Calvin, and Tracy J. Trothen, eds. 2015. *Religion and Transhumanism: The Unknown Future of Human Enhancement*. Santa Barbara: Praeger.

PART I

How Did We Get Here?

CHAPTER 2

From Lamech to Bezalel: Biblical Reflections on the Polyvalence of "Technology"

Steven J. Kraftchick

Introduction

For most of the previous three centuries the term "technology" has commonly referred to "those tools that we see and use," i.e., tools external to our bodies. But the production of new technologies that we not only use but also wear, ingest, or enmesh in our bodies make clear that external and observable tools are only a small part of technology. Technology is not simply an aggregation of the artifacts we use, but our cultural reality. Limiting "technology" to the category of external tools that enhance our control over nature or that expand human capacity precludes us from recognizing its real breadth and influence on society,

This chapter was composed while I was a fellow at the Fox Center for Humanistic Inquiry at Emory University (2015–2016) and I would like to express my appreciation for the opportunities this afforded me. All Scripture quotations are from the New Revised Standard Version.

S. J. Kraftchick (✉)
Candler School of Theology, Emory University, Atlanta, GA, USA
e-mail: skraftc@emory.edu

S. Donaldson, R. Cole-Turner (eds.), *Christian Perspectives on Transhumanism and the Church*, Palgrave Studies in the Future of Humanity and its Successors,
https://doi.org/10.1007/978-3-319-90323-1_2

namely that it is a mode of seeing and existing in the world. To appreciate our actual relationship to technology we must move our attention beyond its pragmatic axis to its interpretive and symbolic functions.[1]

When we reflect on these developments, our awareness of the increasingly blurred boundaries between the "organic person" and the technologically enhanced person cause us to reconsider fundamental questions of our identities. In effect, we realize that we are fast becoming "techno-humans." In some sense this has always been the case, but the current shifts in technological capacities are taking us into a new dimension of this existence. N. K. Hayles has referred to this cultural and biological process as a "'techno-genesis,' a dynamic coevolution between technology and human biological and cultural development."[2]

As a result of this co-evolution, we are also experiencing individual and social disequilibrium. Technologies are not benign; they have both salutary and deleterious effects on us and our world. This amplifies our ambivalence toward technology because we must use and make decisions about individual technologies well before we know what their combined effects may be. Nevertheless, due to their social and moral implications, these decisions must move apace even while we are attempting to interpret the technologies that prompt them.[3]

David Lewin begins his book, *Technology and the Philosophy of Religion*, thusly, "We are surrounded by technologies that have transformed life over the last 100 years. Yet that transformation is far too complex to be considered simply positive or negative, good or bad. So we live with an acute ambivalence towards technology…."[4] Lewin suggests further that, often this ambivalence finds expression in terms of three aspects of our existence, the environmental impact of technological innovations, the physical, social, and psychological effects of technology on ourselves, and our relationships with other beings—human and non-human, and, the

[1] See Winner, *The Whale and the Reactor*, 5–10; Marcia-Anne Dobres, "Meaning in the Making: Agency and the Social Embodiment of Technology and Art," in Schiffer, *Anthropological Perspectives on Technology*. Seminal essays on this matter are, Martin Heidegger, "The Question Concerning Technology," in *The Question Concerning Technology and Other Essays*, 3–35, and Hans Jonas, "Toward a Philosophy of Technology," 34–43.

[2] Hayles, N. Katherine, "Wrestling with Transhumanism," in Hansell and Grassie, *Transhumanism and its Critics*, 216.

[3] See Mary Tiles and Hans Oberdiek, "Conflicting Visions of Technology," in Scharff and Dusek, *Philosophy of Technology*, 249–259. See also, Peter-Paul Verbeek, *Moralizing Technology*.

[4] Lewin, David, *Technology and the Philosophy of Religion*, 4.

possibility that technology will result in nihilism and the spiritual degradation that nihilism entails, a threat that Lewin calls "theological."[5]

Though they are prompted by new developments, our ambivalence toward technology and the questions it raises are not new, but age old.[6] However, two aspects of currently emerging technologies have changed the way we experience this ambivalence and face these questions. First, the possibilities that were once only the stuff of science fiction, are now a part of everyday life. Second, the increasing and rapid relocation of technological adaptation now includes not only the external environment, but also our actual bodies. As a philosophical movement and ethical community, transhumanism has, at its base, moved these possibilities front and center by arguing that technologies from current pharmaceuticals to advanced AI should be used to enhance the human being and reconfigure the external world.[7] It is not too much to say that these two changes strain not only our conceptions of the human being, but also the capacities of our received traditions to engage transhumanists about our imagined futures.[8]

Ron Cole-Turner underscores the implications of this strain, noting that, "We have transhumanists to thank for picking up a word that we were not using (transformation), electrifying it, and now handing it back to us in the form of questions. What is the theological significance of technology, particularly the technology of human enhancement? What is the future of this cosmos, and what role do we humans play in bringing about its future?"[9]

[5] Ibid., 5.

[6] The human-technology intersection has always been a matter of existential concern. Certainly, however, it has become more prominent in the latter half of the twentieth century. See Cuomo, *Technology and Culture in Greek and Roman Antiquity*.

[7] I am aware that there are different forms of transhumanism and that there are indistinct lines between transhumanists and posthumanists. For a discussion of the terminological difficulty see, Hava Tirosh-Samuelson, "Religion," in Ranisch and Sorgner, *Post-Transhumanism*, 49–71.

For the purposes of this paper I am using the term "transhumanism" in the most general way possible. In this regard Nick Bostrom's description is helpful because it sets the claims of transhumanism without overly restricting its varied applications. Bostrom notes, "Transhumanism is a loosely defined movement that has developed gradually over the past two decades, and can be viewed as an outgrowth of secular humanism and the Enlightenment. It holds that current human nature is improvable through the use of applied science and other rational methods, which may make it possible to increase human health-span, extend our intellectual and physical capacities, and give us increased control over our own mental states and moods." Bostrom, "In Defense of Posthuman Dignity," 202–203.

[8] The complexities are helpfully discussed by Allenby and Sarewitz in *The Techno-Human Condition*.

[9] Cole-Turner, "Going Beyond the Human," 22.

Two of Cole-Turner's questions about the theological significance of technology and the role it can play in our mediated futures are of particular interest to us as they pertain to a conversation with transhumanism: the significance of technology and the notion of future agency.[10] To initiate the conversation, this chapter presents some biblical reflections on technology drawn from the books of Genesis and Exodus before reflecting briefly on the eschatological vision found in Revelation 21–22. The chapter then concludes with remarks distilled from Paul's letter to Rome and the Gospel of Matthew. Other important aspects and features of religious belief could be raised in a conversation with transhumanism (e.g., the nature of the human being, the relationship of the self to its body, the active agent in transformation, human-divine synergies, and the difference between immortality as a relationship and individual longevity, among others).[11] However, conceptions of technology as reflections on the relationship of the human to its use of tools and their strategies of employment underlies all of the topics listed above. Indeed, because of its interpretive function, technology engages religion to such a degree that Philip Hefner has suggested, "Everything we think about religion, everything we think is spiritual, is rearranged by technology."[12]

Beginning a Conversation with Transhumanism

While they are not obvious partners, conversations between Christian communities and transhumanists are potentially quite fruitful. First, concerns about the use and function of technologies in and on our lives as well as questions of human value and enhancement are not restricted to those who identify as transhumanists, but are also central to Christian discourse. Second, as we mutually consider the similarities and differences in respective beliefs and worldviews, a deeper understanding of the possibilities of human life and human destiny can occur.

[10] By "conversation" I mean something like Hans-Georg Gadamer's idea that when we are in a conversation, "we have encountered something in the other that we have not encountered in the same way in our own experiences of the world… Where a conversation is successful, something remains for us and something remains in us that has transformed us." In Vessey and Blauwkamp, "Hans Georg-Gadamer: The Incapacity for Conversation (1972)," 355.

[11] I have made an initial attempt to raise these questions in Kraftchick, "Bodies, Selves, and Human Identity," 47–69.

[12] Hefner, *Technology and Human Becoming*, 12.

A conversation such as the one suggested here has the goal of "explication," a term W. V. O. Quine uses to describe a strategy where "we supply lacks. We fix on particular functions of the unclear expression that make it worth troubling about, and then devise a substitute, clear and couched in terms to our own liking, that fills those functions. Beyond those conditions of partial agreement, dictated by our own interests and purposes, any traits of the *explicans* come under the head of 'don't care.'"[13]

Successful explication occurs when conversationalists make concerted efforts to find expressions that move a conversation forward and to let red-herrings swim on. When it comes to complex groups such as religious bodies and the diverse and diffuse transhumanist communities, this can be an especially salutary path to follow because neither of these communities is monolithic nor uses the common terms in the same way. Given this reality, a conversation that seeks explication requires more than a modicum of patience. However, the effort is worth expending since, even if our explications "fail to capture essences or to state analytic truths, [they] tell us how to translate theories from familiar but confusing idioms into idioms better suited to our purposes."[14] Our aim need not be complete agreement, but simply clarification of our aspirations and the means by which we individually and collectively might attain them.

Such outcomes are reasonable goals, but the conversation is important for three other reasons.[15] First, transhumanism's primary foci overlap with those of the Christian tradition on matters such as the qualities of humanity's flourishing, the relationship of the body to the person, who or what is the active agent in achieving that flourishing, and, if one of the goals of human existence is deep personal relationships with others, how this is accomplished and maintained.[16]

[13] Quine, W. V., *Word and Object*, 258–259 as quoted in Jeffrey Stout, "What is the Meaning of a Text?," 2.

[14] Stout, "What is the Meaning of a Text," 2.

[15] My reasoning is similar to William Sweet's in his essay, "Technology, Religion, and Human Destiny," in Feist, Beauvais, and Shukla, *Technology and the Changing Face of Humanity*, 192–204. Sweet notes that technology not only bears on human possibility but also on how we understand human life itself. As these are central to Christian faith a conversation is actually unavoidable.

[16] This is not to say that the two groups agree on the qualities or means to achieve these goals, only that they share interests. On the intersection see, Cole-Turner, *Transhumanism and Transcendence*, and for the issues raised by human enhancement see Savalescu, *Human Enhancement* and Mehlman, *Transhumanist Dreams and Dystopian Nightmares.*

Second, the relationship of human beings to technology is symbiotic and more enmeshed than we usually recognize. Because of its commitments to the combined potentialities of emerging technologies (e.g. genetics, robotics, informatics, nanotechnology [GRIN]) to transform our physical, mental, and social forms of existence, transhumanism makes the contours and depths of this relationship more apparent. Reflecting on the confluence of these advances and the boundaries they blur, transhumanists argue that not only will human lifestyles change, but so will the actual nature and corporeal makeup of human beings. Moreover, since emergent technologies will change the nature of our bodies and our relationship to them, so will the degree of agency and responsibilities we have to advance those changes.[17]

As a result of this, Katherine Hayles has noted that transhumanism, "is fervently trying to figure out where technogenesis is headed in the contemporary era and what it implies for our future. This is its positive contribution and … why it is worth worrying about."[18] Ron Cole-Turner makes Hayles remarks even more pointed in his essay, "Going Beyond the Human: Christians and Other Transhumanists," when he observes that, "Whatever its cause, the rise of secular transhumanism, just now barely credible because of the development of new technologies of human enhancement, must surely *provoke* (my emphasis) Christians to look again with fresh eyes at the core Christian belief and restate the nature of hope that lies at the heart of faith."[19]

Third, these conversations are not just theoretical, they actually shape the way in which we conceive of and conduct our relationship to technology. Because discussions about the role of technologies in our individual and social lives are not merely academic enterprises, but actually concern real changes in our physical beings, our social and economic structures, our relationship to the planet, as well as the lives of future generations, they must be broad in scope and include a range of perspectives. Transhumanists make these dynamics evident, so, to the degree that they

[17] Not surprisingly, these arguments raise questions about the nature of the human being, its ends, and the means by which it attains them, all of which are fundamental to our conceptions of ethics and community. I want to emphasize that these arguments are not raised simply by those who stand outside the transhumanist perspective, but also from those within such as Buchanan, *Better Than Human*, Roden, *Posthuman Life*, and Agar, *Humanity's End*.

[18] Hayles, N. Katherine, "Wrestling with Transhumanism," in Hansell and Grassie, *Transhumanism and its Critics*, 216.

[19] Cole-Turner, "Going Beyond the Human," 20–26.

help us think clearly about future realities and our role in bringing them about, we should participate actively in a dialogue with them.[20]

This conversation is already well underway among philosophers, scientists, doctors, engineers, poets, novelists, and artists, who are all engaged in discussions about the issues which accompany our emerging transhumanism.[21] Historically, religious communities have been the locus for these conversations, and they should occupy this space again in order to bring other important ideas and voices to the conversation.

One final caveat before entering the discussion. The biblical texts quite obviously do not speak directly about such things as nanotechnology, stem cell research, or computer informatics. The biblical writers could not imagine our mundane contemporary technologies, let alone the GRIN technologies that are being developed for immediate future use, so there is no need to advocate a simple extraction of biblical texts. However, the biblical texts can supply criteria to evaluate technology when it is understood as a mode of viewing, encountering, and mediating the world. To appropriate scripture in a manner that engages this more accurate conception of technology requires the extrapolation of first principles from the texts rather than an extraction of specific directives from them. Carefully read scripture becomes a catalyst and nexus for our thinking and our conversations, as well as a place where our questions are more carefully crafted (and sometimes changed) rather than peremptorily answered.[22]

Scriptural Reflections on Technology

Technology in Genesis 3–6

Given the pervasiveness of technology, our ambivalence toward it, and Hefner's observation that it is rearranging our thoughts about religion, it is reasonable to see how our ancestors reflected on the relationship of

[20] As with many people these are personal as well as intellectual matters for me. Over the last three years five of my colleagues, friends, and family have been diagnosed with life threatening illnesses. None of these people were or are egregious abusers of their bodies, indeed one is an infant. The ethics and ethical use of technologies requires serious conversations with other people who are searching to understand our future relationships to these mediating tools and to technology as a mode of existence. See, for example, Hughes, *Citizen Cyborg*.

[21] See for only one example, Schneider, *Science Fiction and Philosophy*.

[22] See on the different modes of Scripture reading, Morgan and Barton, *Biblical Interpretation*, and on the role of theology as communal deliberation, Tanner, *Theories of Culture*.

technology to their religion. A practical beginning point is found in Genesis 3:21, "And the Lord God made garments of skins for the man and his wife, and clothed them." Recall that Genesis 3 records the disobedience to God's injunction not to eat from the tree of the knowledge of good and evil (2:17) and the devastating results of that act: the curse upon the earth, the animals, and the people who were enjoined to populate it (3:17). However, immediately after these pronouncements, Genesis recounts that God made garments for Adam and Eve. The first narrative event after the pronouncement of punishment is an event of grace, eventuated through technological means.[23]

Technology becomes central to the narrative in the next chapter where Abel is characterized as a keeper of sheep and Cain a tiller of the ground. This is followed by a listing of Cain's descendants, that includes Enoch who built a city, and Lamech, whose sons Jabal and Jubal are referred to as the ancestors of those "who live in tents and have livestock" and "who play the lyre and the pipe" respectively. Lamech also fathered Tubal-cain who "made all kinds of bronze and iron tools" (4:17–22). These technologies are not explicitly depicted as good or bad, simply as part of human life. However, when Lamech proclaims that "I have killed a man for wounding me, a young man for striking me. If Cain is avenged sevenfold, Lamech is seventy-seven fold," (4:23b–34), the association of the technologies with Lamech's name suggests that technologies also have a potential for danger and damage.

By the sixth chapter of the Genesis narrative human society has devolved to the point that God is sorry to have made humans and determines to eliminate humans from the creation. God thus announces to Noah that, "I have determined to make an end to all flesh, for the earth is filled with violence because of them" (6:13). Given the remarks about Lamech and his ancestors in chapter 4, a reader could easily infer that the use of

[23] Note that God's garments made from animal skins replace the temporary "fig leaf" garments (Gen 3:7) made by the primal couple. Even when the human technology fails, the divine gift can redeem the situation. In contrast to the majority of ancient texts that treat the origins of human technologies as a divine gift, the Genesis account focuses on their human invention. This is not to say that the divine agent never introduces technology in Genesis, only that technology is understood as a source of both the progression and regression of humankind. See, Robert Di Vito, "The Demarcation of Divine and Human Realms in Genesis 2–11," in Clifford and Collins, *Creation in the Biblical Tradition*, 39–56 and David P. Melvin, "Divine Mediation and the Rise of Civilization in Mesopotamian Literature and in Genesis 1–11," 1–15.

technology is involved in this violence, or that technology is the means of destruction, if not its source. However, as in chapter 3, immediately after the pronouncement God orders Noah to "make yourself an ark of cypress wood, make rooms in the ark, and cover it with pitch. This is how you are to make it: the length of the ark three hundred cubits, its width fifty cubits, and its height thirty cubits." In other words, once again, God's redemptive action is instantiated through technological means. As 1 Peter suggests, the ark is another sign of God's redemptive purposes, for through it and its sojourn on the water humankind was saved (1 Peter 3:20).

Much more might be made of these texts, but here we can simply note that read in isolation from chapters 3 and 6, the characterization of technology in chapter 4 would suggest a negative evaluation. However, when the texts are read as part of a narrative, it becomes clear that technology is also seen as a locus of redemption. On the one hand, the texts argue that, wrongly employed, technology breaks relationships. On the other, they also portray technology as a vehicle of the divine/human that enables us to move toward full relationships with ourselves, with the earth, and eventually with the Creator. The narrative thus shows that technologies create vectors of intent, which humans have a responsibility to discern.

Technology in Exodus 25–40

Perhaps because of the influence of films like *The Ten Commandments* or *The Prince of Egypt* (themselves technological marvels) or more recently, Ridley Scott's *Exodus: Gods and Kings*,, we often remember Exodus as the book of Israel's release from Egyptian bondage, the wandering in the Wilderness, or the covenant established at Sinai. These are major events in Exodus, but they comprise only the first twenty-four chapters of the text. However, with the exception of chapters 32–34, which recount the worship of the Golden Calf, the final third of Exodus is devoted to a detailed architectural and engineering description of the Tabernacle, the ark of the covenant, and the furnishings and implements used in the sanctuary—in other words, a technological model and its construction. In effect, Exodus presents as much technological material as it does "salvific" narrative, suggesting that the two are inseparably entwined. In fact, Exodus 25:8–9 makes just that point when Yahweh instructs Moses to have the Israelites "make me a sanctuary, so that I may dwell among them. In accordance with all that I show you concerning the pattern of the tabernacle and all its furnishings, so you shall make it." Thus, Exodus conjoins technology with the presence of God in Israel's midst. To put it succinctly, no technology, no presence of Yahweh—a breach of the fundamental relationship that worship of Yahweh should signify.

The verses that precede this pronouncement detail what is entailed in fulfilling it: "The Lord said to Moses: Tell the Israelites to take for me an offering; from all whose hearts prompt them to give you shall receive the offering for me. This is the offering that you shall receive from them: gold, silver, and bronze, blue, purple, and crimson yarns and fine linen, goat's hair, tanned ram's skins, fine leather, acacia wood, oil for the lamps, spices for the anointing oil and for the fragrant incense, onyx stones and gems to be set in the ephod and breastplate." (25:1–7).

Note that all of the metals, the gems, the material goods, the potions, and the garments are products of technology—in some cases very complex technology. Without these capacities and production, the Tabernacle cannot be constructed and if it cannot be constructed, the commands of Yahweh will not be met, preventing the abiding of Yahweh among them.

Following this introduction, three chapters are devoted to specific instructions for constructing the ark, the table of presence, the lampstand, the Tabernacle itself (including its framework and curtains), the altar, the court and its curtains, and the oil for the lamps. Another two provide directions for the manufacture of the priests' garments and procedures for ritual sacrifice (chapters 28 and 29). The entire section is composed of seven speeches by Yahweh: (1) this extended speech (25:1–30:10); (2) the institution of a census tax, which requires a technology of accounting, "designated for the service of the tent of meeting," and as a reminder of the ransom given for Israel (30:11–16); (3) instructions for manufacturing the bronze ablution bowl for ritual cleansing (30:17–21); (4) instructions for creating the oils and compounds needed for anointing (30:22–33); (5) a recipe for the composition of the incense used in worship (30:34–38); (6) the designation of Bezalel as the chief engineer and artisan who would (along with Oholiab) "make all that [Yahweh] has commanded," (31:1–11), and (7) a command that Moses instruct the people to keep the Sabbath (31:12–17) which serves as the capstone for this section.

These descriptions are repeated in chapters 35–39, allowing the author to show how obedience is a process of moving from vision to implementation. To hear of the tabernacle is not enough, and to build it without attention to its purpose would be a form of apostasy, as chapters 32–34 suggest when the pseudo-object of worship (the calf) is constructed. To bring their faith to fruition, the Israelites must listen to God and use their technological capacities to carry out his commands.

A significant shift in the ordering of these instructions occurs in chapters 35–39. In the first recitation, the command to observe the Sabbath comes at the end of the instructions. In the second instance, it is at the head of the list, effectively placing Sabbath observance at the center of the instructions and their execution. Sabbath, or observance of proper time, is enabled by its own technology, but the Sabbath synchronizes how and when we will use our technologies. As God states in 31:13, "You shall keep my sabbaths, for this is a sign between you throughout your generations, given in order that you may know that I, the Lord, sanctify you." The seriousness of this command is underscored when Moses repeats it in 35:2 adding to it the severe penalty of death for those that break the command. The Sabbath is not an option therefore, but a necessary part of existence with Yahweh. By creating a hiatus of time and human effort, the possibility of proper relationship with Yahweh is restored, pointing toward an ultimate Sabbath or reconciliation.[24] As we will see, this theme reappears in the vision scene from Revelation.

Before leaving this rich text, let us note two of its other features. The first was alluded to above in reference to construction of the golden calf. The Exodus composer seems to have deliberately placed the account of Israel's apostasy (chapters 32–34) between the two descriptions of the Tabernacle plans and its construction. The framing thus creates a contrast of technology's capacities for good and bad. In the middle chapters, technology appears when the calf is formed of smelted gold and cast in a mold (32:4) to create a simulacrum of Yahweh's mercy seat (cf. 25:17–22). In the chapters that flank this episode, technology is a means of mediating God's presence. The irony is strong here—when Israel uses its technological capacity to create an idol it devolves quickly into a mass of undifferentiated people. The narrative suggests that when factors of fear and isolation arise, human can use technologies as their means of restoration or trust. As in Genesis, the humans have broken their relationship with God, and the makeup of God's people is threatened with dissolution. In contrast, when the peoples' technologies are employed for the purposes of maintaining relationships with God, they are avenues for flourishing and fulfillment of religious confessions.

[24] Further connections between the Genesis creation account and Exodus 25–31 are considered by Levenson in *Creation and the Persistence of Evil*, 82–86.

This is underscored by the designation of Bezalel as the Tabernacle's chief artisan. His name means "in the shadow of God" or more directly the one who is within God's protection and benevolence. According to 31:3 God has filled him with "divine spirit," with ability, intelligence, and knowledge in every kind of craft. The first phrase "divine spirit" (***ruah elohim***) is also used in Genesis 1:2 where the spirit of God moves over the face of the waters, creating order from chaos. Bezalel, through his technological gifts and knowledge and endowed with this same spirit creates another form of order, a space where God's presence can again be revealed in the world and dwell.

As William Propp suggests, the term "divine spirit" usually applies to an ability to interpret dreams or to a prophetic announcement—both theological enterprises. Here, however, it connotes wondrous mental and manual skill, "the ability to actualize divine intent."[25] Bezalel's technological skill, as a demonstration of his "divine spirit," thus serves as a means for interpreting divine intent—not, as is typically the case, through oral announcement, but through the production of material goods. In effect the technological artisan becomes a theologian "exampling divine activity and rendering it active and comprehensible."[26] Waldemar Janzen's exposition of this text makes this point in another way. He states, "In other words, 'spiritual gifts' are not reserved here for the realm with which we often associate them (e.g. prayers, prophesy, etc.). Instead they are applied to the work of artists and artisans working with tangible materials. [T]he term 'incarnational' seems appropriate; God works through the earthly, bodily, and material function of human beings."[27]

This brief review of these texts does not pretend to solve our ambivalence about "technology," but to show that such ambivalence is unavoidable. The texts also underscore that technology belongs not only to a secular set of pursuits or a realm outside of the sacred, but is also as engaged with one's religious actions and beliefs as any other aspect of our existence.

Technology in Revelation 21–22

The eschatological portraiture of the New Testament is better conceived as a pointillist painting than a set of blueprints for the end times and one should take seriously Jesus' riposte that no one knows the time or manner

[25] Propp, *Exodus 19–40*, 487.
[26] Ibid.
[27] Janzen, *Exodus*, 368.

of its advent. There are some features of eschatological existence that seem clear, however. First, according to Paul, eschatological existence is embodied existence, albeit with a body that differs from the one we currently have (1 Cor 15:35–49). Human identity requires not only spirit, but embodiment. Second, eschatological existence is impervious to the vagaries of finitude. Third, the present is constructed in light of the future rather than the present determining the future.[28] Finally, while this world is passing away, it is to be redeemed—granting value to the world even if it is not yet perfected. The New Testament eschatological stance is simultaneously a critique of all claims to perfection through human efforts *and* a call to live and act with reconciliation toward the world. Numerous texts reflect these stances, but I have chosen Revelation 21–22, the vision of the New Jerusalem, for comment.

The passage is not chosen randomly. Revelation 21:1–22:5 recounts a vision of the New Jerusalem given to the prophet John during a mystic experience. This vision follows the revelation of the defeat of Satan, Death, and Hades (20:7–15) and the dawning of a new heaven and a new earth. In this final vision the prophet sees a new Jerusalem descending "out of heaven from God" (21:1–2). John then hears a voice from the throne of heaven announcing that the home (Greek: *skēnos* = tabernacle) of God is with mortals and that God will dwell (Greek: *skēnoō* = to tabernacle) with them—"they will be his peoples, and God himself will be with them" (21:3). The use of the term "tabernacle" to describe God's presence with human beings is an allusion to Ezekiel 37: 26–28 and the vision of the Valley of the Dry Bones, when God will restore Israel to its full complement.[29] However, the use of the term also brings to mind God's injunction to Moses to have Israel construct a dwelling place, the tabernacle (Exodus 25:8–9).

The foundation stones identified in Revelation 21:19–20 offer additional links to the Exodus materials. First, the types of stones recall those embedded in the priest's breastplate (Exodus 28:17–21) and second, the breastplate is inscribed with names of the twelve tribes of Israel (28:21),

[28] See on this Burdett, *Eschatology and the Technological Future.*

[29] Revelation 21–22 follows a pattern found in Ezekiel 40–48 where the Temple is restored along with Jerusalem. However, the Ezekiel narrative is a rehearsal of the Exodus materials since the Tabernacle is a prototype for the Temple. In effect, the author of Revelation has appropriated the Temple/Jerusalem traditions to make them universal. Cf. Prigent, *Commentary on the Apocalypse of John,* 597.

which in Rev. 21:12 are inscribed on the new city's gates.[30] Through this combination the author signifies that the new reality of God's peoples incorporates all humanity—Israel *and* the nations. That is, the reconciliation of the world's peoples occurs through the act of God in Christ.[31]

Revelation 21:9–22:9 is constructed as a parallel to the description of Rome in chapter 17:1–19:10. The two cities are set in contrast, the earthly city only a parodic facsimile of the heavenly reality. One is the epitome of human endeavor and technological marvels but lives in isolation from God, convinced of its own power. The other city exists through the initiative and grace of God, given to people for their constant and continual existence with God. God's city is described as "a bride adorned for her husband;" Rome as "the great whore set on many waters." Rome is destroyed, becoming a haunt in which demons dwell (18:3) because, rather than worship God, the nations have become drunk and committed fornication with her (17:2). The New Jerusalem descends from heaven providing the spring water of life (21:6; 22:1–3) and a place where the nations walk by the light of God's glory (21:23–24). Instead of the worship of the technologies of industry and commerce (18:11–24), there is continual worship directly in the presence of God (22:3–5).

The contrast between Rome's idolatry and the true worship in the new dwelling mirrors the contrast between the false worship of the calf in Exodus and the proper devotion to God that the tabernacle will facilitate. This link is suggested by the phrasing in Rev 17:2, 3; 18:3, 9; 19:2, which refers to the "fornication" that the nations have committed with Rome. Similar language is used in Exodus 32:6, where we are told that the people "sat down to eat and drink and rose up to revel." Rome is also adorned with gold, jewels, and pearls (18:16), but this wealth is destroyed "in one hour" (18:17). Just as in the Exodus narrative where the calf idol is destroyed and then replaced by the Tabernacle, so Rome, the false city, is replaced by the new Jerusalem, the true dwelling place of God.[32]

The New Jerusalem is still a technological marvel as the description of its dimensions and structures suggest, but, as a gift of God, it is not made by human hands nor is it an attempt to displace God as the locus of devo-

[30] See on the allusion, Prigent, *Commentary on the Apocalypse of John,* 617, and Aune, *Revelation 17–22,* 1187.

[31] Cf. Ephesians 2:11–22 where similar imagery is used to illustrate that the people of God include both Jew and Gentile and that together they form the Temple where God dwells.

[32] The last term in Exodus 32:6—"revel"—is suggestive of orgiastic activity as Paul indicates when he used the incident as an example to warn the Corinthians against idolatry (1 Cor 10:7) See Moberly, *At The Mountain of God,* 46.

tion. Symbolically the author reminds the readers that ultimate salvation is not achieved by human ingenuity or technical competence, no matter how great, but through the redemptive actions of God on behalf of people.

In concluding the treatment of this passage, allow me to comment on a surprising feature of this text, namely what the prophet did *not* see in this vision. According to 21:22 John "saw no temple in the city, for its temple is the Lord God the Almighty and the Lamb." Given the tenor of Revelation this is a surprising omission, but it is deliberate. We noted in our discussion of Exodus that technology was a necessary means of mediation in and with the world and in relationship with God. In the New Jerusalem, that is in the newly formed relationship with God, this mediation is not needed. This is emphasized through the doublet to verse 22, where the New Jerusalem is described as a state where there is "no need for sun or moon to shine on it, for the glory of God is its light, and its lamp is the Lamb" (21:23). In the Genesis account the sun and moon are created in order to create "time" (Gen 1:14–18). In the New Jerusalem time does not exist, showing that eternity is not an extension of the time-space continuum; it is beyond or other than such realities. In effect, technology is only of use this side of eternity. It is and will remain a constant of our historical existence but, in the consummate existence of relationship with God, it is rendered penultimate and ceases to have such a function. The use of the phrase "face to face" in Rev 22:3–4 implies a direct relationship with God, hence, one with no need for mediation. The reason there is no temple in the New Jerusalem is that this technological creation is no longer necessary. The relationship of God to humans is made immediate and direct—that is to say, it has been restored and redeemed.[33]

Conclusion

Given these biblical reflections on the role and nature of technology, let us return to the possibility of a conversation with transhumanism concerning the force and goals of our own engagements with technology. We need not claim that theologians must become transhumanists in the modern sense of that term, or that all transhumanists must be interested in exploring "non-science" ethics questions for a successful conversation to

[33] This is the logic behind Paul's comments in 2 Cor 3:18. Christ, the "image of God" (2 Cor 4:4) reflects the glory of God, transforming humans into the same image "from one degree of glory to another."

occur. Rather, it is sufficient for us only to investigate our respective understandings of humans as technological beings who are in transformation. Should that happen, each interested party can address the fundamental questions of technology's roles for societal well-being, the limits of those roles, and our own roles as responsible agents in transforming both ourselves and our society.

My goal has been to show that the Christian tradition has resources for addressing these questions—resources that are meaningful to all parties involved. Thoughtfully engaged, they can prompt thick conversations of explication about the human being's interaction with technology in all its complexities, and so provide a baseline for addressing questions such as the means and ends for individual and corporate human enhancement, particularly how we determine our role in constructing the future.

It is important, however, to remember that incorporating biblical materials into these conversations requires proper respect for their overall perspectives. Rarely do the biblical texts speak of technology in a direct manner and this is especially true of emerging technologies. They speak more directly to the role of the human being in the construction and use of technology. Thus, when the biblical materials speak of technology they allow remarkable freedoms but also constant warnings about misplaced confidence. Further, unlike most transhumanist materials, when the biblical sources speak of human enhancement, they speak not only of physical change, but personal relationships with other beings and with God. In these texts enhancement through technology does not refer to individual perfection, but to communal wholeness. The Christian claim of resurrection focuses on the power and willingness of God to restore and maintain life and relationship against all forces. As we saw in the texts from Revelation, the restoration and redemption of all human beings remains the *telos* of God's activity in the world (Cf. Romans 8:37–39). Further, Rev 21–22 suggest that eschatological existence is not confined to the human or physical sphere. To be sure the created order is restored to its proper relationship with God, but as Romans 8:19–21states, "the creation waits with eager longing for the revealing of the children of God, for the creation was subjected to futility, not of its own will but by the will of the one who subjected it, in hope that the creation itself will be set free from its bondage to decay and will obtain the freedom of glory of the children of God." That restoration entails reunification with the divine being that brought it into existence (1 Cor 15:20–28), implying that extended physical life, renewed natural resources, reconstructed neurons, muscles, and cells, as marvelous as they are, remain elements of this life, not the restoration to fully realized existence, which the New Testament attests.

Given such an orientation the question becomes how one should live toward it. This is the essential question of Christian ethics. In an adumbrated form some possibilities arise when we consider the criteria by which those who claim allegiance to Jesus are evaluated in Matthew 25 when all of humanity will face the end of human history. Recall that neither the sheep nor the goats know themselves as such. It is only when the Son of Man reveals their identities that they are made aware of their true selves. Surprisingly the criteria for this revelation are not enumerated in terms of doctrine or creed, but rather with regard to actions on behalf of the hungry, thirsty, those without shelter, those who were sick, and the strangers among us. Here we see some of the most significant ends of religious allegiance to Jesus, the alleviation of human suffering and deprivation.

The biblical texts show that technology is inherently ambiguous and that our relationship to it is always an ambivalent one. This cannot be avoided, though it may be reduced. Technology is part of the human condition, so it cannot solve that condition. It can, however, alleviate some of its effects. The texts also suggest some ways to assess our integration with technology in this regard. Ultimately this is a matter of two things: a deep recognition that we are part of an ecology and our stewardship of that ecology. The matter here is not one of fear, at least not fear of extending beyond our limits. As the texts show, it is not invention or even the use of technology that dictates its value, but the goals desired in that use. Inherently the human being is a techno-being, shaping and shaped by its technology—in fact, this is one way to define us. However, as the texts suggest, technology does not in and of itself define us. The relationship is not deterministic, but symbiotic. Thus, as with all the gifts of existence, material and spiritual, the evaluative question is one answered in terms of intent. In the end, the question is not simply what is the relationship of the human being to technology, but rather, what is the relationship to God that the techno-human reflects in its behavior and hopes. This, of course, is never answered solely in relationship to an individual in isolation, but rather in the relationships of humans to other persons. That is to say, the human being is a person in relationship to other persons, past, present and future. Indeed, at the most microscopic of levels, the human is a person in relationship to the most basic matter of the universe. The human is therefore, kin to all of creation.

For this reason, our evaluations of technology and transhumanism should be measured primarily in terms of sustaining and enhancing that relationship. To the degree that the transhumanist commitment to technological development aids in the alleviation of thirst, hunger, poverty, and sickness among the least of us, we should welcome its overtures. To the degree that it conceives of the human as a relative to other creatures and their environments, we should heed its reflections. On the other hand, to the extent that technological mediation leads to the dehumanization of members of the human community or earth's inhabitants, we should call it into question. When it conceives of local and individual enrichments of human capacities as an enhancement of "all of humankind and the creation" we must demur and remember that such claims of power have rarely been as extensive as their advocates suggest and almost always come at a cost to someone or something else. None of our claims to self-sufficiency are unalloyed. In every case, it appears that the engagement depends neither on technological facility nor on doctrinal purity or group identities, but on the capacity to identify and enter into relationship with the "invisible Christs" among us. As at the beginning of human history, so also until its end, Scripture suggests that our technologies and our relationship to them will be judged by their service to this primary goal.

Bibliography

Agar, Nicholas. 2013. *Humanity's End: Why We Should Reject Radical Enhancement*. Cambridge, MA: MIT Press.

Allenby, Braden, and Daniel Sarewitz. 2011. *The Techno-Human Condition*. Cambridge, MA: MIT Press.

Aune, David E. 1997. *Revelation 17–22 Volume 52C: Word Biblical Commentary*. Waco: Word.

Bostrom, Nick. 2005. In Defense of Posthuman Dignity. *Bioethics* 19 (3): 202–214.

Buchanan, Allan. 2011. *Better Than Human: The Promise and Peril of Enhancing Ourselves*. New York: Oxford University Press.

Burdett, Michael S. 2015. *Eschatology and the Technological Future*. New York: Routledge.

Clifford, Richard J., and John J. Collins, eds. 1992. *Creation in the Biblical Tradition*. Washington, DC: Catholic Biblical Association.

Cole-Turner, Ronald, ed. 2011. *Transhumanism and Transcendence: Christian Hope in an Age of Technological Enhancement*. Washington, DC: Georgetown University Press.

———. 2015. Going Beyond the Human: Christians and Other Transhumanists. *Dialog: A Journal of Theology* 54 (1): 20–26.

Cuomo, S. 2007. *Technology and Culture in Greek and Roman Antiquity*. New York: Cambridge University Press.

Feist, Richard, Chantal Beauvais, and Rajesh Shukla. 2010. *Technology and the Changing Face of Humanity*. Ottawa: University of Ottawa Press.

Hansell, Gregory R., and William Grassie, eds. 2011. *H+/−: Transhumanism and Its Critics*. Philadelphia: Metanexus Institute.

Hefner, Philip. 2003. *Technology and Human Becoming*. Minneapolis: Fortress.

Heidegger, Martin. 1977. *The Question Concerning Technology and Other Essays*. New York: Harper and Row.

Hughes, James. 2004. *Citizen Cyborg*. Boulder: Westview.

Janzen, Waldemar. 2000. *Exodus*. Scottdale: Herald.

Jonas, Hans. 1979. Toward a Philosophy of Technology. *Hastings Center Report* 9 (1): 34–43.

Kraftchick, Steven John. 2015. Bodies, Selves, and Human Identity: A Conversation Between Transhumanism and the Apostle Paul. *Theology Today* 72 (1): 47–69.

Levenson, Jon. 1988. *Creation and the Persistence of Evil*. San Francisco: Harper and Row.

Lewin, David. 2012. *Technology and the Philosophy of Religion*. Newcastle upon Tyne: Cambridge Scholars.

Mehlman, Maxwell. 2012. *Transhumanist Dreams and Dystopian Nightmares*. Baltimore: Johns Hopkins University Press.

Melvin, David P. 2010. Divine Mediation and the Rise of Civilization in Mesopotamian Literature and in Genesis 1–11. *Journal of Hebrew Scriptures* 10: 1–15.

Moberly, R.W.L. 1983. *At the Mountain of God: Story and Theology in Exodus 32–34*. Sheffield: JSOT.

Morgan, Robert, and John Barton. 1988. *Biblical Interpretation*. Oxford: Oxford University Press.

Prigent, Pierre. 2001. *Commentary on the Apocalypse of John*. Trans. Wendy Pradels. Tübingen: Mohr Siebeck.

Propp, William H.C. 2006. *Exodus 19–40*. New Haven: Yale University Press.

Ranisch, Robert, and Stefan Sorgner, eds. 2014. *Post-Transhumanism: An Introduction*. New York: Peter Lang.

Roden, David. 2015. *Posthuman Life: Philosophy at the Edge of the Human*. New York: Routledge.

Savalescu, Julian. 2009. *Human Enhancement*. Oxford: Oxford University Press.

Scharff, Robert C., and Val Dusek. 2014. *Philosophy of Technology: The Technological Condition an Anthology*. 2nd ed. Chichester: Wiley.

Schiffer, Michael Brown, ed. 2001. *Anthropological Perspectives on Technology*. Albuquerque: University of New Mexico Press.

Schneider, Susan, ed. 2016. *Science Fiction and Philosophy: From Time Travel to Superintelligence*. 2nd ed. Hoboken: Wiley.
Stout, Jeffrey. 1982. What Is the Meaning of a Text? *New Literary History* 14 (1): 1–12.
Tanner, Kathryn. 1997. *Theories of Culture: A New Agenda for Theology*. Minneapolis: Augsburg Press.
Verbeek, Peter-Paul. 2011. *Moralizing Technology: Understanding and Designing the Morality of Things*. Chicago: University of Chicago Press.
Vessey, David, and Chris Blauwkamp. 2006. Hans Georg-Gadamer: The Incapacity for Conversation (1972). *Continental Philosophy Review* 39 (4): 351–359.
Winner, Langdon. 1986. *The Whale and the Reactor: A Search for Limits in an Age of High Technology*. Chicago: University of Chicago Press.

CHAPTER 3

Dragons and Dog-Headed Saints: Some Medieval Perspectives on the Significance of the Human Form

Allison Hepola

Introduction

Transhumanism, especially in its post-humanist forms, forces us to confront the possibility of someday encountering what Nick Bostrom describes as, "beings whose basic capacities so radically exceed those of present humans as to be no longer unambiguously human by our current standards."[1] Needless to say, this poses a number of significant questions and challenges for Christianity. However, transhumanism would not exist as a viable philosophy or movement were it not for specific scientific and technological advancements that only occurred in the last sixty years. Given this it might seem strange to consider medieval perspectives as we explore the relationship between transhumanism and Christianity. Here I will put forth the

[1] Cole-Turner, "Introduction: The Transhumanist Challenge," in Cole-Turner, *Transhumanism and Transcendence*, 13.

A. Hepola (✉)
Department of Philosophy, Rutgers University, New Brunswick, NJ, USA

S. Donaldson, R. Cole-Turner (eds.), *Christian Perspectives on Transhumanism and the Church*, Palgrave Studies in the Future of Humanity and its Successors,
https://doi.org/10.1007/978-3-319-90323-1_3

position that there are in fact valuable insights to be found in certain medieval texts that can inform and enrich a contemporary Christian scholar's approach to transhumanism. While obviously the medievals were unaware of things like genetics, nanotechnology, and cyborgs, some of them did believe in the existence (or at least took it to be a legitimate possibility) of something else: person-like creatures inhabiting faraway lands with bodies that differed radically from standard human anatomy. These are what St. Augustine terms the "monstrous races" in *City of God*. The monstrous races were also beings that were not "unambiguously human," and for the medievals, these races were not just mere curiosities. Their potential existence posed serious and challenging questions for Christianity. The medievals were interested in questions like: Are the monstrous races human? Are they part of God's plan or a deviation from it? Are they subject to Original Sin? Should they be evangelized, baptized, and brought into the church? I believe the insights of the medievals on the monstrous races can inform and enrich contemporary Christian discussions of transhumanism as we, like the medievals, ponder the relationship between human nature and the appearance and operations of the body, the theological significance of radically different human bodies, and the ethical implications of attempts to transcend bodily limitations.

The Monstrous Races

The monstrous races have a long history. They are a familiar feature of Greco-Roman mythology where we encounter cyclops, satyrs, centaurs, gorgons, and a host of other creatures that are humanlike yet inhabit bodies very different from those of ordinary humans. Outside of the classical world, we find monstrous races in the mythological traditions of Northern Europe where giants, dwarves, leprechauns, elves, and similar creatures abound.

But the monstrous races were not confined to myth and legend. The foremost scientific authority of the ancient world, Pliny the Elder, discusses the monstrous races at length in Book VII of his *Natural History*. Pliny describes races of cannibals, pygmies, giants, hermaphrodites as well as some more flamboyant creatures. There were the Blemmyae, a group of people who have no heads and necks at all. Instead their faces are on their torsos, with the eyes just below the shoulders.[2] There were the Sciopods

[2] Pliny, *Natural History Books 3–7*, Book VII. II. 21–24.

who hopped about on a single leg with one extremely long foot on the end. Sciopods were said to lie on their backs with their feet in the air in order to shade themselves from the midday sun.[3] There were the Cynocephali, people of great height with human bodies but the heads of dogs.[4] According to Pliny, these monstrous races lived in the furthest reaches of Africa, India, and the Caucuses—far away from where any Roman could expect to travel.

Pliny's treatment of the monstrous races gave them a good deal of intellectual respectability which would persist well into the Middle Ages. It is impossible to know exactly just how widespread belief in the monstrous races was in the Middle Ages. But we do know some things about the monstrous races' place in medieval thought. While the monstrous races were never a central concern for anyone, the subject was remarked upon by major intellectuals like Augustine, Isidore of Seville, Albert the Great, and several scholars at the School of Paris. Texts referring to the monstrous races appear in nearly every corner of Christendom, from Ethiopia to Iceland. Artistic depictions of the monstrous races can be found in illuminated manuscripts, maps, icons, and architectural details.[5] Interest in the monstrous races did not wane until the Age of Discovery.

Are They Human?

The question of whether the monstrous races were bona fide humans was especially important for medieval Christians. The question was significant for Christians in a way it was not for the Greeks and Romans. For medieval Christians to ask if some being is a human is not merely a question of whether we are applying the right categories to the natural world (although that certainly was of interest to many medievals!). It is also to ask if the monstrous races are implicated in original sin, if Christ's redemption applies to them, if they are in need of evangelization and baptism.

Not surprisingly the answer to the question of the monstrous races' humanity that had the greatest influence on medieval thought was St. Augustine's. In Book XVI of *City of God*, Augustine briefly discusses the monstrous races. He mentions Cyclops, pygmies, hermaphrodites,

[3] Ibid.
[4] Ibid.
[5] Friedman, *The Monstrous Races in Medieval Art and Thought*, 37–58; 131–162.

Sciopods, Blemmyae, Cynocephali, among others.[6] At the beginning of his discussion, Augustine expresses reservations that any of these beings described by pagan authorities actually exist, and he stresses that Christians are not required to believe in them.[7] Probably because of this skepticism, Augustine takes no firm position on whether the monstrous races are human or not.

But Augustine does allow for the possibility that the monstrous races exist and are humans. He puts forth two necessary conditions for a creature to count as human. The creature must be rational and it must descend from Adam.[8] As long as these conditions are met, then it does not matter what the creature's body looks like. Augustine writes:

> No faithful Christian should doubt that anyone who is born anywhere as a man – that is, a rational and mortal being – derives from that one first-created human being. And this is true, however extraordinary such a creature may appear to our senses in bodily shape, in color, or motion, or utterance, or in any natural endowment, or part, or quality.[9]

To support this, Augustine goes on to discuss how occasionally human infants are born with birth defects that make them appear very different from their parents. Yet no Christian doubts that these infants are human and descend from Adam with all that implies theologically.[10] Likewise, Augustine writes, there could be entire groups of people that appear very different from ordinary humans yet this would not mean they are not human and are not descended from Adam. In other words, for Augustine the appearance of the body is not particularly connected to human nature.

Ultimately though, Augustine does not provide definitive answers to the major questions surrounding the monstrous races. He acknowledges that it is possible that (assuming they exist) Blemmyae, Cynocephali, and the like are humans. But it is also possible that such creatures exist but are not humans. Augustine offers his readers no means of determining which possibility actually is the case. But this task would be taken up by other medieval thinkers.

[6] Augustine, *City of God*, Book XVI, Chapter 8, 661–662.
[7] Ibid., 662.
[8] Ibid.
[9] Ibid.
[10] Ibid., 663.

One noteworthy contribution comes from a Carolingian-era scholar, Ratramnus of Corbie. In a letter to a friend, Ratramnus discusses the Cynocephali.[11] He acknowledges that when we first hear tales of Cynocephali, our first inclination is to doubt the humanity of these creatures because they have animal heads and bark instead of speaking, speech being an essential characteristic of humanity. But then Ratramnus points out that according to the stories of Cynocephali that he and his friend were familiar with

> They form a society and live in villages; they cultivate fields and harvest crops; they cover their private parts through human modesty rather than exposing them like beasts; and for garments they use not merely skins but true clothes by which they indicate their modesty. All of this leads you to believe that they possess a rational soul.[12]

For Ratramnus, social behavior is the mark of rationality and thus of humanity. If the accounts of Cynocephali engaging in rational, political behavior are correct, then we have every reason—as far as Ratramnus is concerned—to believe that the Cynocephali are humans. The bizarre appearance of the Cynocephali is irrelevant. Like Augustine, Ratramnus believes that the outward appearance of the body has little bearing on whether or not a creature counts as human.

Another interesting argument in favor of the humanity of the monstrous races comes from the thirteenth-century Franciscan scholar Alexander of Hales.[13] In a work titled *Summa Universae Theologiae*, Alexander says that Cynocephali, Blemmyae, and other monstrous races are indeed human and descend from Adam. Their strange appearance is due to the Fall. Animals who cannot sin (because they are not rational) also cannot suffer the "disordering effects of sin" in their bodies.[14] Only rational creatures—that is, only humans—can suffer the consequences of the Fall in a bodily way. Since groups like the Blemmyae bear the disordering effects of sin in their bodies, Alexander reasons, this proves that they must be human and not animals.

Not every medieval who discussed the monstrous races was convinced of their humanity. Some thought that the monstrous races could simply be exotic animals like apes that had been badly described by

[11] Friedman, *The Monstrous Races in Medieval Art and Thought*, 188.

[12] Ibid.

[13] Ibid., 187.

[14] Ibid.

exaggeration-prone travelers.[15] But even if the monstrous races existed exactly as they were described, that did not necessarily mean they were human. One ground for denying the humanity of the monstrous races was that their bodies deviated too much from ordinary humans'. We see this line of thought developed by a thirteenth-century Scholastic named Peter of Auvergne. In a *quodlibet* concerning the pygmies, Peter argues that the bodies of humans, like all other animals, are subject to limits on their size.[16] To be a human is to be no smaller than some size and no larger than some other size. According to Peter, the average human height is four cubits. Giants are said to be twelve or more cubits tall, three times as tall as the average human. But pygmies are said to be only half a cubit tall, eight times shorter than the average human. Pygmies deviate much more from the average human height than giants, therefore pygmies are not humans, Peter concludes (it is not clear if Peter thought giants were human).

The political life of the monstrous races used by Ratramnus to support their humanity was also disputed by other medievals. In his commentary on Aristotle's *De Animalibus*, Albert the Great discusses pygmies.[17] Albert appears to approach the subject with an unargued-for assumption that the pygmies are not human to begin with. He goes on to explain how all their apparently human behavior—speech, social life, the use of tools, etc.—is just an imitation of what they observe real humans doing, analogous to the behavior of trained animals.

I do think it is noteworthy that explicit arguments against the humanity of the monstrous races are rather weak and do not come until relatively late in the Middle Ages. Overall it seems that when the topic of the monstrous races was considered by the medievals, the tendency was to view them as humans. Christian thinkers in this period were surprisingly comfortable with the claim that the appearance and structure of the body is largely irrelevant to human nature. Instead the most important feature of the body for determining whether some being is human or not is simply whether that body biologically descends from Adam. I believe this insight from the medievals is useful to keep in mind as contemporary Christians consider transhumanism and post-humanism which offer the possibility of humans whose bodies might end up looking and operating as differently from standard humans' as the bodies of the Blemmyae and Cynocephali.

[15] Ibid., 25.
[16] Ibid., 195.
[17] Ibid., 191–192.

God's Curse and God's Plan

But before we can further explore the connections between medieval thought on the monstrous races and contemporary Christian perspectives on transhumanism, I must address a major disanalogy. Transhumanism is, of course, about improving, exceeding, and transcending the human condition. The radically different beings envisioned by Bostrom and others are meant to be *better* in all sorts of ways than ordinary, unenhanced humans. Meanwhile the monstrous races are *degenerate* and, well, *monstrous.* They are *worse* than ordinary humans in all sorts of ways. No one wants to become a Sciopod or Cynocephalus.

The degenerate status of the monstrous races was certainly recognized by the medievals. It was especially emphasized in the mythological and folkloric traditions concerning the monstrous races. There are a number of tales in Anglo-Saxon, Celtic, German, and Jewish mythology where the monstrous races are said to be descendants of Cain (occasionally Noah's disobedient son Ham is the ancestor of the monsters).[18] The monstrous appearance of these races is a direct result of God's curse of Cain. Probably the most famous example of this tradition is the villain Grendel from *Beowulf* of whom we learn:

> He had dwelt for a time
> in misery among the banished monsters,
> Cain's clan, whom the Creator had outlawed
> and condemned as outcasts. For the killing of Abel
> the Eternal Lord had exacted a price:
> Cain got not good from committing that murder
> because the Almighty made him anathema
> and out of the curse of his exile there sprang
> ogres and elves and evil phantoms
> and the giants too who strove with God[19]

In the more scholarly medieval treatment of the monstrous races we see a less polemical attitude. Augustine, for instance, views the monstrous races as playing some mysterious role in God's overall design of the universe. He cautions readers against questioning God's intentions in allowing the monstrous races to exist, writing

[18] Ibid., 93–107.
[19] *Beowulf*, lines 104–113.

> But if we assume that the subjects of those remarkable accounts are in fact men, it may be suggested that God decided to create some races in this way, so that we should not suppose that the wisdom with which he fashions the physical being of men has gone astray in the case of the monsters which are bound to be born among us of human parents; for that would be to regard the works of God's wisdom as the products of an imperfectly skilled craftsman.[20]

Yet while Augustine does not attribute the monstrous races to a divine curse and acknowledges that God has his reasons for creating them, they are still *monstrous* races. They are not something that any normal-bodied human would want to become. And this presents a challenge for making connections between medieval ideas about the monstrous races and contemporary transhumanism. Nonetheless, I believe there is still more to be learned from the monstrous races and their place in medieval thought. I will now turn my attention to two medieval accounts of monstrous races that involve an improvement (of sorts) of the monstrous condition. These tales can give us a more detailed picture of medieval attitudes toward human nature and the significance of the body's appearance. This can, in turn, enrich our contemporary considerations of transhumanism.

Two Tales of Transcendence

The first tale is the Icelandic *Volsunga Saga*, written by an unknown author in the thirteenth century. It is not a wholly Christian text, to be sure; Odin, Loki, and other figures from paganism feature prominently in the narrative. But like all Norse literature from this time period, the *Volsunga Saga* has a foot in each world. The author was Christian and could not have helped but view and interpret the oral tradition of the Norse through a Christianized lens.[21] I have chosen to focus on it because it involves a monstrous race—the dwarves—but also includes elements of transcendence.

Near the beginning of the saga, we are introduced to the king of the dwarves Hreidmar who has three sons: Otr, Regin, and Fafnir. Otr has the magical ability to shape-shift into an otter. Fafnir is described from the outset as fierce and greedy.[22] One day while Otr is in the form of an otter,

[20] Augustine, *The City of God*, 663–664.
[21] Byock, *The Saga of the Volsungs*, 7–8.
[22] Ibid., 57.

Loki throws a stone at him and kills him. In Norse culture the punishment for murderers is the payment of a *weregild*—literally "man gold"—to the family of the deceased. Loki brings a huge amount of gold and treasure to Hreidmar, but not before placing a curse on it that it "would be the death of whoever owned it."[23] Soon Fafnir, being greedy, kills his father and takes the treasure all for himself, leaving his surviving brother Regin with nothing. Regin approaches Sigurd—the human hero of the saga—for help. He tells Sigurd about the remarkable thing that happened to Fafnir:

> Fafnir became so ill-natured that he set out for the wilds and allowed no one to enjoy the treasure but himself. He has since become the most evil serpent and lies now upon this hoard.[24]

Over an indeterminate period of time, Fafnir has transformed into a large and fearsome dragon, also known as a "worm." He now lives alone in a lair at the top of a thirty-fathom high cliff. He spends his days in solitude, lying on his vast hoard of golden treasure. By the way, if this sounds familiar, it should; the tale of Fafnir was one of Tolkien's inspirations for the characters of both Smaug and Gollum.

Sigurd the human eventually makes his way to Fafnir's isolated den and deals the dragon a mortal wound. As Fafnir is dying, he converses with Sigurd, having retained the power of speech. When the dragon finally dies, Sigurd cuts out his heart, roasts it, and eats it. This gives Sigurd the power to understand the speech of birds.[25]

There are a couple of significant things going on in this story. First, notice that the tale of Fafnir is one of transcendence. He begins his life as a dwarf. Through his transformation, he reaches and then exceeds the bodily capabilities of not just dwarves but of humans too. He grows to a great size, he can fly, and he can breathe poison. Fafnir becomes the most physically powerful and terrifying creature in the world of the saga. It is difficult to know what to make of the magical powers of comprehension eating Fafnir's heart gives to Sigurd, but perhaps it could imply that Fafnir's intellectual capacities were also enhanced by his transformation.

Secondly, the way in which Fafnir's transformation came about is interesting. It is quite different from the human-to-monster transformations

[23] Ibid., 58.
[24] Ibid., 59.
[25] Ibid., 66.

we find in other mythological sources like Ovid where a human's body is changed through the explicit and direct action of a god. In Fafnir's case, the gods play little role. The gold Fafnir acquires is indeed cursed, but the curse of Loki only entails death—not bodily transformation—to those who possess it. Both Hreidmar and Sigurd are killed after receiving the cursed treasure, but neither undergoes any sort of bodily transformation.

Instead the reasons for Fafnir's transformation are more ambiguous. The transformation seems to proceed from a character trait, greed that Fafnir already has long before the cursed gold comes to his family. It is perhaps no coincidence that greedy Fafnir transforms into a dragon instead of something else like an ogre. In the blend of Christianity and paganism that was the medieval Norse mindset, dragons were symbolic of the sin of greed.[26] While the cursed gold no doubt played some role in Fafnir's transformation, the saga indicates there may be another cause—Fafnir's greed itself. Fafnir's sin deformed, distorted, and dehumanized him.

It should be noted that many of the authors of the Norse sagas studied in Paris—at the height of the Scholastic era—before returning to Iceland to compose their texts. Given that, it is perhaps not unreasonable to detect some natural law themes in the story of Fafnir. Human nature is teleological—to be human is not simply a matter of having a body arranged in the right sort of way. To be human is to have a function, an inherent ordering toward an end—toward virtue. The more we pursue virtue, the more human we become. Conversely the more we pursue vice, the less human we become. This is literally the case with Fafnir. Even though Fafnir's physical and mental capacities far outshine those of humans and dwarves, he has not transcended anything. He has degenerated from what he ought to have been due to his unrestrained greed.

The next tale I would like to discuss involves one of the most obscure and enchanting legends of the Middle Ages, the dog-headed saint. Stories about the conversion of Cynocephali are believed to have originated in the East in the earliest centuries of the church. For reasons that remain mysterious, the stories of the Cynocephalic saint migrated to the West where they became associated with the legend of St. Christopher.[27] This version of the St. Christopher legend seems to have always been a fringe phenomenon, but it may have left its mark on the more mainstream

[26] Tolkien, J.R.R. "Beowulf: The Monsters and the Critics." in Tolkien *The Monsters and the Critics and Other Essays*, 17–23.

[27] White, *Myths of the Dog-Man*, 34–36.

Golden Legend where St. Christopher is described as extremely tall with a fearsomely ugly (though otherwise human) face.[28]

Here I will refer to an account of a Cynocephalus saint from *The Contendings of the Apostles*, a fourteenth-century Ethiopian text that is believed to be a translation of a now-lost fifth-century Greek, and possibly Nestorian, text.[29] At the beginning of the narrative, the apostles Andrew and Bartholomew are in Parthia attempting to evangelize a hostile city. Dwelling outside the city walls is a man-eating Cynocephalus named Abominable. He is very tall and fearsome in appearance. He cannot speak like a human, but instead barks like a dog. One day an angel appears to Abominable with a message from God: he is to seek out the apostles and accompany them into the city. When he hears this, Abominable expresses his concerns to God, saying:

> O my Lord, I am not like all other men, for my face is not like that of a man, and I have no knowledge of their speech. Now, if I go with them, where shall I find food? And if I be hungry, where shall I find men to eat? I should certainly then fall upon them and devour them.[30]

But the angel reassures Abominable, telling him

> "God will give unto thee the nature of the children of men, and He will restrain in thee the nature of the beasts." And in that same hour the angel stretched out his hands and brought out the man with a face like unto that of a dog from the fire, and he made over him the sign of the cross, and cried out unto him in the Name of the Father, and of the Son, and of the Holy Ghost. Then straightaway did the nature of the beast go forth and out of him, and he became as gentle as a lamb.[31]

Abominable goes forth and soon encounters Andrew and Bartholomew. They are terrified by his monstrous appearance, but having obtained the ability to speak like a human, Abominable calms them by saying, "be not afraid, O my spiritual fathers."[32] The apostles are eventually convinced that Abominable has been sent to them by God and soon they all enter the city together.

[28] Voragine, *The Golden Legend*, 173.
[29] White, *Myths of the Dog-Man*, 25.
[30] Budge, *The Contendings of the Apostles*, 205.
[31] Ibid., 205–206.
[32] Ibid., 207.

Not long after this, governor of the city arrests Andrew and Bartholomew and sentences them to be eaten alive by wild beasts. To save the lives of the apostles, Abominable prays that God will turn him "back again into [his] former nature."[33] The prayer is answered and all of Abominable's gentleness departs. "He leaped upon all the wild beasts that were among the multitudes of people who were gathered together, and he slew them forthwith, and tore out their bowels and devoured their flesh."[34] This spectacle convinces the governor and all the people in the city to repent and do homage to the apostles. Andrew and Bartholomew pray over Abominable and "in that same hour the nature of the children of men returned unto him, and he became as gentle as a lamb."[35] He is given a new name, "Christian." The story ends with the baptism of the entire city, but it is not clear if Abominable himself is baptized.

As in the *Volsunga Saga*, this tale depicts a member of a monstrous race who transcends his condition. But unlike Fafnir, Abominable does not undergo a change in the appearance and operation of his body. When the angel's blessing gives Abominable a human nature, the main effect of this is gentleness. He no longer has an insatiable appetite for human flesh. He loses his propensity to violence; this is vividly displayed when Abominable asks God to temporarily remove his human nature to fight the beasts. But very importantly, Abominable's physical appearance is not changed. He retains his dog head and remains so horrifying looking that the apostles faint from fear when he first approaches them. Even at the close of the narrative when the whole city is baptized and Abominable is given his new name, he remains a Cynocephalus.

Yet thanks to the power of God, Abominable has become a human. To be human is not a matter of inhabiting a body of a certain sort; it is a matter of disposition, of virtue. Like the *Volsunga Saga*, the story of Abominable offers us a teleological view of human nature where sin makes us less than human, virtue makes us truly human, and the appearance and capabilities of our bodies are largely irrelevant. Fafnir, despite the great physical powers he gains from his transformation into a dragon, has undergone degeneration. Abominable, despite the great ugliness of his body that is left unchanged, has through the grace of God transcended the limits of his condition in a way Fafnir could have never dreamed of.

[33] Ibid., 210.
[34] Ibid.
[35] Ibid., 212.

Conclusion

As Christian scholars today ponder what genetic engineering, artificial intelligence, and other bodily enhancements imply for human nature, we can draw upon insights gained from medieval discussions of the humanity of the monstrous races. While the possibility of "no longer unambiguously human" beings presented to us by transhumanism seems frighteningly unprecedented, we can take heart in the fact that this is not actually a new issue for the church—Christians have grappled with it before in the form of the monstrous races.

Going forward, I believe there are a few specific ways in which insights from the medievals can inform and enrich contemporary Christian understandings of transhumanism. We learn from studying the medieval engagement with the monstrous races that some Christians were surprisingly comfortable with allowing creatures as bizarre in body as Cynocephali and Blemmyae to count as human. Perhaps this insight could alleviate concerns on the part of contemporary Christians that the radical changes to the body's appearance and operations proposed by some transhumanists would result in the destruction of our shared human nature.

At the same time, we learn from the medievals that while the appearance of a creature's body may be somewhat irrelevant, whether or not a creature descends from Adam matters quite a lot for its counting as a human. This insight might lead us today to place more significance on how closely a post-human creature is biologically or genetically connected to the rest of the human species. Perhaps we may find that we need to approach the issue of humans who have undergone genetic enhancements quite differently from how we approach the issue of artificial intelligence.

But most importantly of all, the medieval engagement with the monstrous races reminds us not to overlook the connection between human nature and virtue as we consider the challenges raised by transhumanism. Stories like those of Fafnir and Abominable raise the point that true transcendence might mean more than simply overcoming the limits of the body and the mind.[36]

[36] I would like to thank Andy Poker and S. Kate Devitt for their helpful suggestions and feedback on an earlier version of this chapter.

Bibliography

Augustine. 2003. *City of God*. Trans. Henry Bettenson. London: Penguin Classics.

Beowulf. 2000. Trans. Seamus Heaney. New York: W.W. Norton.

Budge, E. A. Wallis (translator). 1935. *The Contendings of the Apostles*. London: Oxford University Press.

Byock, Jesse (author and translator). 1990. *The Saga of the Volsungs*. London: Penguin Classics.

Cole-Turner, Ronald, ed. 2011. *Transhumanism and Transcendence: Christian Hope in an Age of Technological Enhancement*. Washington, DC: Georgetown University Press.

de Voragine, Jacobus. 1998. *The Golden Legend*. Trans. Christopher Stace. London: Penguin Classics.

Friedman, John Block. 2000. *The Monstrous Races in Medieval Art and Thought*. Syracuse: Syracuse University Press.

Pliny. 1947. *Natural History Books 3–7*. Trans. H. Rackham. Cambridge: Harvard University Press.

Tolkien, Christopher, ed. 1983. *The Monsters and the Critics and Other Essays*. London: Harper Collins.

White, David Gordon. 1991. *Myths of the Dog-Man*. Chicago: University of Chicago Press.

CHAPTER 4

"When That Which Is Perfect Is Come": Henry Drummond and "The Changed Life"

Carol Ann Vaughn Cross

Introduction

For life, with all its yields of joy and woe,
And hope and fear…
Is just our chance o' the prize of learning love,
How love might be, hath been indeed, and is.[1]
Robert Browning

Attention to spirituality is one of the most important challenges of a transhumanist society. If transhumanism is the "power of technology to transform humanity" beyond existing physical and intellectual limitations,[2] then we arguably already live in a transhumanist—or at least transhumanizing—society.

[1] Browning, "A Death in the Desert" in *Dramatis Personae*.

[2] Cole-Turner, "Introduction: The Transhumanist Challenge" in Cole-Turner, *Transhumanism and Transcendence*, 4.

C. A. Vaughn Cross (✉)
Samford University, Birmingham, AL, USA
e-mail: cavaughn@samford.edu

S. Donaldson, R. Cole-Turner (eds.), *Christian Perspectives on Transhumanism and the Church*, Palgrave Studies in the Future of Humanity and its Successors,
https://doi.org/10.1007/978-3-319-90323-1_4

Our enmeshment with our technology is part of our human identity. We off-load our memories and other mental functions, and we re-wire our expectations. We live with technologically mediated experiences that influence our perceptions of ourselves, others, and God. As Ted Peters observes, "Technique has not only expanded our practical lives; it has also entered into our inner lives." Our "self-understanding as human beings" involves a "technological mindset" that many people take for granted.[3] It matters the extent to which we consciously and unconsciously use our technology (and for what purposes), or if we allow our technology to use us. Unexamined passivity can have significant moral consequences. Thus our attention to the care and orientation of our inmost being is critical for humanity. This was a concern of at least one nineteenth-century science enthusiast.

An almost universally popular Christian figure, Henry Drummond (1851–1897) concluded that changes in the human condition brought about by our scientific explorations and technological innovations need not be debasing, demoralizing, or dehumanizing if we give proper attention to our inmost beings. Ordained in the Church of Scotland and employed for 19 years as a natural science lecturer at Glasgow's Free Church College, the erudite clergyman earned a significant place in the history of religious interpretations of transhumanism, influencing such later philosophers as Pierre Teilhard de Chardin and Pitirim Sorokin.[4] His scientifically informed theological anthropology addressed critical issues of time, memory, relationships, and moral progress. His popular enchiridion on *The Greatest Thing in the World* rearticulated for modern audiences a classic devotional practice of the Two Great Commands to Love and was cherished by scores of Christians and non-Christians alike, including Christian revival leader Dwight L. Moody and Hindu political leader Mohandas Gandhi, who followed Drummond's "program" in his own practice of *ahimsa*—compassion or humane behavior—as he faced incivility, brutality, and potential destruction.[5]

[3] Peters, "Progress and Provolution: Will Transhumanism Leave Sin Behind?" in Cole-Turner, *Transhumanism and Transcendence*, 75–76.

[4] King, "Love—A Higher Form of Human Energy" 77; David Grumett, "Transformation and the End of Enhancement: Insights from Pierre Teilhard de Chardin" in Cole-Turner, *Transhumanism and Transcendence*, 37–49; Sorokin, *The Ways and Power of Love*.

[5] Emilsen, "Gandhi, Scripture, and the Bible," 83–84.

Drummond embodied the spirit of interdisciplinarity among the sciences and humanities rather than over-specialization, and his model serves us well in our attempts to give meaning to human experience. He was exceptionally skilled at communicating his synthesis of Neo-Platonism, evolutionary theory, Christian transcendentalism, Eastern Orthodoxy, and the Protestant "Higher Life" movement to disparate groups. Drummond presented complex theories in practical terms that general audiences could understand at multiple levels.[6] The "key to all Christ's teaching," he observed, was that "He spoke not to the reason but to the imagination."[7] Seeking to be a "prophet, poet, and scientist" like those he admired, Drummond wove quotes of Goethe, Hugo, Ruskin, Eliot, and the Brownings to engage audiences' scientific and moral imaginations.[8] An unusually unifying figure, Drummond was popular with Christians of widely different denominations, new converts, adherents of other religions, and agnostics. Generations after he died, individuals referred to Drummond as one of "the most Christ-like" humans they had met, a person who "lived in" 1 Corinthians 13.[9] Indeed, his central thesis that Love is life's purpose, "energy," and Greatest Good dominated all of his scientific and theological works.

The Greatest Good

Drummond posited that humanity's past, present, and future can be distilled into one great question, and he directed all of his work to it: "What is the *summum bonum*—the supreme good?"[10] Regarding science and technology, Drummond seemed to have subscribed to Kranzberg's first "law of technology" a century before Kranzberg articulated it. This principle states that our technology is neither good nor bad, nor is it neutral.[11]

[6] R., H. d., "Henry Drummond," in Maine, *The Greatest Thing in the World*, 5–6; Corts, *Henry Drummond: A Perpetual Benediction*, xviii; Drummond, *Natural Law in the Spiritual World*, 376.

[7] Drummond, "Clairvoyance" in Maine, *The Greatest Thing in the World by Henry Drummond*, 189.

[8] Drummond, *Natural Law in the Spiritual World*, 5–22.

[9] Smith, *The Life of Henry Drummond*, 8–9; Corts, *Henry Drummond: A Perpetual Benediction*, xviii; Corts and Corts, "The Man Who 'Lived in First Corinthians 13'" in *The Greatest Thing in the World*, Samford edition, 22.

[10] Drummond, *The Greatest Thing in the World*, Samford edition, 27–29.

[11] Kranzberg, "Technology and History," 544–560.

We choose what we do with our technology, and that determines whether it is a force for Good or not.[12] It is therefore imperative that we pay more, rather than less, attention to the human will.

Drummond approached the Greatest Good and human will in a Classical as well as Christian sense. Every person, he maintained with Plato and Aristotle, is interested in improvement in some way. All humans seek their own version of *eudaemonia*: the good or happy life. The Sermon on the Mount addresses how such happiness, or blessing, works in human lives, culminating in Jesus' command to "be perfect … as your heavenly Father is perfect."[13] In *Natural Law in the Spiritual World*, *Ascent of Man*, *The Greatest Thing in the World*, and "The Changed Life," Drummond addressed this "infinite desirability, the infinite difficulty of being good."[14] In his address on "Christianity and Evolution" at the historic 1893 World Parliament of Religions at Chicago's World Fair, Drummond proclaimed that the "object of Christianity" is "the making of higher and better" individuals "in a higher and better world"[15] because such Goodness—understood as the Perfect Character of Love—"is what we are made for."[16] Humans can discern and participate in this ultimate Good in this life. And if the Church is not committed to this activity, Drummond warned, Christianity is nothing more than "the world's conundrum."[17]

One reason for the conundrum, Drummond noted, is that many religious people assume that the Greatest Good is faith. Many people identify themselves primarily as "people of faith" and discuss science and technology from "faith-based" perspectives. Drummond declared such unexamined assumptions "wrong" and akin to sin because they "miss the mark."[18] In every one of his numerous works, Drummond asserted that the Greatest Good, the ultimate law, and the supreme force of life is Love. He emphasized it in his first best-seller by quoting Robert Browning: "I spoke as I

[12] Bimber, "Three Faces of Technological Determinism" in Smith and Marx, *Does Technology Drive History?*, 80–100.

[13] Matthew 5:48.

[14] Drummond, "The Changed Life" in Maine, *The Greatest Thing in the World by Henry Drummond*, 98.

[15] Drummond, "Christianity and Evolution," 783.

[16] Drummond, "The Changed Life" in Maine, *The Greatest Thing in the World by Henry Drummond*, 98.

[17] Ibid., 98–99.

[18] Drummond, *The Greatest Thing in the World*, Samford edition, 27–29.

saw. I report, as a man may of God's work—*all's Love, yet all's Law.*"[19] He concluded his last publication with 2 John: "the end of Man is 'that which we had from the beginning, that we *love*.'"[20] And he dedicated *The Greatest Thing in the World* to the Perfect Character found in 1 Corinthians 13.[21] Like his Russian contemporary Nicolai Fedorovich Fedorov, Drummond believed that the laws of nature are "rooted in love."[22] All of nature, including humans, exists because of this Love and are subject to its omnipotence. We are born from it and for it.

Drummond observed that our desire for longer life and immortality is "bound up with love." Echoing classical wisdom, he wrote: "We want to live for ever for the same reason that we want to live tomorrow. Why do we want to live tomorrow? It is because there is some one who loves you, and whom you want to see tomorrow, and be with, and love back. There is no other reason why we should live on than that we love and are beloved."[23] Well-versed in the ontological, theological, and teleological questions humans have, Drummond's conclusion and central message were simple: "To love abundantly is to live abundantly."[24]

Such abundance, according to Drummond, occurs because Love is a spiritual and physiological "energy" that nurtures, sustains, and regenerates us.[25] This energy is the "basis of life" that "already be existing" in each of our souls.[26] Like light, humans cannot originate it.[27] And like any energy, Love never dies. It can only be transformed.[28] Drummond reasoned that because Love is an energy, according to God's natural laws, we humans cannot destroy it. Love, Drummond opined, "is something more than all its elements—a palpitating, quivering, sensitive, living thing" that animates physical creatures.[29]

[19] Drummond, *Natural Law in the Spiritual World*, 22, quoting Robert Browning (*Saul*, 35).

[20] 2 John 1:5–6; Drummond, *Ascent of Man*, 345–46.

[21] Drummond, *The Greatest Thing in the World*.

[22] Burdett, "Transcendence and Human Enhancement" in Cole-Turner, *Transhumanism and Transcendence*, 28, 31.

[23] Drummond, *The Greatest Thing in the World*, Samford edition, 61–62.

[24] Ibid., 61.

[25] Drummond, *Natural Law in the Spiritual World*, 289.

[26] Ibid.

[27] Drummond, *The Greatest Thing in the World*, Samford edition, 51.

[28] Drummond, *Natural Law in the Spiritual World*, 289.

[29] Drummond, *The Greatest Thing in the World*, Samford edition, 51.

Drummond's theory of Love as an energy working according to natural laws included a declaration that "metempsychosis is a fact."[30] Metempsychosis in Drummond's works, as in John Donne's poetry, refers to souls' movement, or transmigration, between material bodies.[31] Drummond believed that an omnipotent Creator could include metempsychosis in the natural laws and that Christian traditions recognized it, citing Harvard professor Francis Bowen and his essay "Christian Metempsychosis."[32] "Love not only was; it is; moves; it spreads," Drummond wrote in *Ascent of Man*.[33] He frequently quoted Elizabeth Barrett Browning and Goethe to illustrate spiritual "face-to-face" knowledge and soul-to-soul mirroring that humans experience. In *Natural Law in the Spiritual World*, he asserted that "Nature and man can only form and transform. Hence when a new animal is made, no new clay is made. Life merely enters into already existing matter, assimilates more of the same sort and rebuilds it."[34] The Greatest Law therefore is that Love is an eternal energy that never dies; it innervates and transcends material bodies, and it defies human understanding of time and place. According to Drummond's explication, Love is the energy that draws people together into "transcendent living" through what he called mirroring, involution, and advolution.[35]

Drummond concluded that the Greatest Good occurs when humans fully express the Two Great Commands: the Jewish Shema's Love of God and the Golden Rule's Love of others.[36] According to Drummond, when "That Which Is Perfect is come"[37] into our lives we act more fully in the disciplines of Love, which he enumerated as nine qualities: Patience, Kindness, Generosity, Humility, Courtesy, Unselfishness, Good Temper, Guilelessness, and Sincerity.[38] We can become habituated to asking ourselves two questions of every decision and action: *Does this*

[30] Drummond, "The Changed Life" in Maine, *The Greatest Thing in the World by Henry Drummond*, 105.

[31] "Transmigration of Souls."

[32] Drummond, "Christianity and Evolution."

[33] Drummond, *Ascent of Man*, 345.

[34] Drummond, *Natural Law in the Spiritual World*, 228–89.

[35] Drummond, *The Greatest Thing*, Samford edition, 51. David Grumett, "Transformation and the End of Enhancement: Insights from Pierre Teilhard de Chardin" in Cole-Turner, *Transhumanism and Transcendence*, 46–47.

[36] Deuteronomy 6:4–5; Leviticus 19:18; Matthew 22:37; Mark 20:30; Luke 10:27.

[37] 1 Corinthians 13:10, King James Version.

[38] Drummond, *The Greatest Thing in the World*, following 1 Corinthians 13:4–7.

improve myself in God's eyes? Does this benefit or diminish Love in others? In Drummond's schemata of "The Changed Life" this is the only standard by which any choice we make or any action we take can be considered improvement or progress.

Perfect Character Versus Self-Improvement

Drummond maintained that humans can use scientific and technological developments to pursue the Greatest Good, serving God and humanity. Like many transhumanists, Drummond believed that a scientific understanding of the "most mystical point" of life can help us realize a "better" humanity.[39] He saw a divine command for us to use our scientific knowledge and technology to find this "mystical point" of life, which Drummond variously called the "Nobler Form," "The Perfect Type," "Christ Consciousness," and "Glory." This Perfect Character or "most mystical point" of life Drummond considered to be both fully human and fully divine, the hypostatic model of Christ wherein full humanity is fully united with the divine. We pray for this when we pray God's "will be done, on earth as it is heaven."[40]

According to Drummond, the Perfect Character of 1 Cor. 13:4–7 is the model and energy force that transforms our self-centered struggles into the Greatest Good. Emphasizing hope for a better future and faith in our progress toward it, Drummond confidently preached that "That Which Is Perfect" can be found in this life more than many people initially believe. "[T]o enter Heaven a man must take it with him,"[41] he stressed. And he noted that we will only pursue such Character if we believe it is attainable. Due to this "Image of Christ that is forming within us"[42] Christians of all people should be optimistic rather than pessimistic about the future, Drummond asserted. He agreed with other nineteenth-century Christian intellectuals who believed in the inevitability of progress as they approached the last century of the second millennium.[43] And like many of his

[39] Hardesty, *Faith Cure.*

[40] Matthew 6:10.

[41] Drummond, "Clairvoyance," in Maine, *The Greatest Thing in the World by Henry Drummond*, 189. Drummond, *The Greatest Thing in the World*, 45.

[42] Drummond, "The Changed, Life" in Maine, *The Greatest Thing in the World by Henry Drummond*, 116.

[43] Drummond, "Christianity and Evolution," 782–3.

contemporaries, including some missionaries, Drummond advocated using all tools of science and technology to bring about a literal union of heaven and earth wherein God's will and human will are fully united.[44]

Drummond valued scientific knowledge as a means of drawing closer to God and evangelized for it accordingly: "The new energies in the world demand a will and an ever present will. To science God no longer made the world and then withdrew; He pervades the whole. Under the old view God was a non-resident God and an occasional wonder worker. Now He is always here."[45] A popularizer of Charles Darwin's evolutionary theories among Christian clergy and laity,[46] Drummond enthused that evolution would be "the greatest generalisation the world has ever known" because it affects every "category of thought."[47] He believed fervently that scientific advances could lead to theological ones, rather than the other way around. Using himself as an example, Drummond demonstrated how scientific knowledge can affect our ontological questions and improve our theological doctrines of revelation, immortality, the Holy Spirit, and sin.[48] He suggested that "if science can help us in any way to know how sin came into the world, it may help us better know how to get it out."[49] Like Bacon centuries earlier, Drummond believed that refusal to allow scientific progress to inform our theology is spiritually lazy. Science "has given us a more Godlike God,"[50] he exulted. "Plato's prisoner has his 'face to the light.'"[51] Thus, "the whole question of the incarnation is beginning to assume a fresh development."[52]

The point of such pursuits is not self-aggrandizement but the inculcation of the Perfect Character for the Greatest Good. Whereas some transhumanists today advocate "the making of higher and better" humanity through individual "Technological Self-Transformation,"[53] Drummond

[44] Richard, *New Testament of Higher Buddhism*; Hardesty, *Faith Cure*; Timothy George, Foreword, in *The Greatest Thing in the World*, Samford edition, vii.

[45] Drummond, "Christianity and Evolution," 782.

[46] Drummond, "Christianity and Evolution;" Lightman, "Darwin and the Popularization of Evolution," 17–19.

[47] Drummond, "Christianity and Evolution," 782–3.

[48] Ibid.

[49] Drummond, "Christianity and Evolution," 782–3; Drummond, *Natural Law in the Spiritual World*, 385.

[50] Drummond, "Christianity and Evolution," 782.

[51] Drummond, *Natural Law in the Spiritual World*, 47.

[52] Drummond, "Christianity and Evolution," 782–3.

[53] More, "Technological Transformation."

presented a "larger view" of progress in which the Perfect Character is exercised and shared, not hoarded. His society was as enamored of the idea of self-improvement as ours is. From Socrates to Descartes to today, Western culture has been obsessed with the improvement of the existential self. An individual's perception of the self in Drummond's lifetime, as in ours, could change drastically with scientific discoveries and technological innovations. New communication technologies, for example, transgressed previous limits of time and space and opened greater possibilities for visible and invisible connections. Increasingly addicted to technologies of speed and convenience, modern individuals struggled to distinguish between luxuries and necessities for their existence.[54] Despite his own enthusiasm for progress, Drummond cautioned against the competition, impatience, isolation, and indolence we can cultivate with spiritually destitute uses of science and technology.

Although the term was not coined until the late twentieth century, Drummond was aware of what is called self-fetishization. He warned of its challenges, especially for zealous Christians.[55] In consumer cultures humans can both "consume" and "become" material objects, objectifying and dehumanizing themselves and others. With our technology, we can turn the individual self into a "project" and a "product" physically, socially, and spiritually. This process becomes easier and faster the more advanced our technology becomes. We are often unaware that we are doing it; the less we pay attention, the less we distinguish our technological tools and cultural forces from our own will. In "The Changed Life," Drummond pointed out ways that religious people can fetishize spiritual progress. His contemporaries might spend a lot of time or money experimenting with new mechanical gadgets, electrical novelties, and pharmaceuticals intended to offer short cuts to insight, productivity, or well-being. Such spiritual self-techniques, Drummond wrote, can become narcissistic and dangerously self-obsessed.[56] Without regularly pausing to examine our inner selves and the Perfect Character, we end up in self-loathing, self-aggrandizement, or self-righteousness.

In the West, as our material conveniences increase, so do our expectations of instant gratification and quick fixes. The more accustomed we

[54] Kern, *The Culture of Time and Space*; Marvin, *When Old Technologies Were New*.

[55] Ewen, *Captains of Consciousness*, 47.

[56] Drummond, "The Changed Life" in Maine, *The Greatest Thing in the World by Henry Drummond*, 98–100.

become to faster connectivity, the more we raise expectations and demand escalation. Our impatience with the status quo grows, and we seek greater shortcuts. As Drummond noted, these addictions to speed and convenience can afflict our inner being. When we measure progress by immediate results, our impatience with ourselves, others, or God increases. This cycle of impatience can lead us to use our science and technology in ways that fall short of the Greatest Good, especially if we are not aware of it in the first place. Drummond drew parallels between our "hurry up" attitude toward God and our impatience with nature.[57] One of Drummond's major themes was that all progress, including the spiritual kind, occurs according to God's natural laws of evolution, not in spite of them. In his best-selling *Natural Law in the Spiritual World,* Drummond affirmed that nature does not make leaps, and we should calibrate our expectations accordingly.[58] Just as change in God's natural world does not occur randomly or haphazardly, neither does spiritual progress.[59] "The Christian Life," he wrote, "is not a vague effort after righteousness—an ill-defined pointless struggle for an ill-defined pointless end. Religion is no disheveled mass of aspiration, prayer, and faith."[60]

Playing with multiple meanings of the term "kingdom" scientifically, geographically, and allegorically, Drummond described how visible and invisible kingdoms grow according to natural laws of generation, degeneration, and regeneration: "rising tier above tier in ever increasing sublimity and beauty, their foundations visibly fixed in the past, their progress … [are] the signs … that 'the Kingdom of God is at hand.'"[61] According to Drummond's Neo-Platonic and Christian transcendentalist metaphysics, progress is not strictly linear but rather an evolutionary dialectic of forward/upward and backward/downward movements—with a momentum ultimately to/back to the Source of life. According to Drummond's theories of energy, if what transhumanists today call a singularity occurs, such

[57] Drummond, *Natural Law in the Spiritual World.*

[58] Ibid., 6. Drummond paraphrases Charles Darwin (*The Origin of Species*, 223), "Natura non facit saltum," who is quoting Carl Linnaeus, *Philosophia Botanica*, 40 (i.e., "nature does not make leaps").

[59] Drummond seems to have interpreted "apparent randomness" as our inability to see the actual order of everything from the divine point of view or "big picture." (cf. *Natural Laws in the Spiritual World*).

[60] Drummond, *Natural Law in the Spiritual World*, 285.

[61] Ibid., 391.

a post-human state is the same as the pre-human state of union with the Creator.[62] Drummond therefore did not question the inevitably of "progress" understood in this way.

While some religious individuals and transhumanist enthusiasts believe that the only way to improve humanity is to diminish and eventually eradicate the human will, Drummond rejected that notion because he was convinced that God's laws—natural and spiritual—are omnipotent and rooted in Love. Our humanity, he insisted, is God's creation from Love, in Love, and to Love. Thus the human will is not lost in the process of Perfection, as it is re-united with the Perfect Character of Love. It is transformed by and into the Perfect Character. It is this divinely created human will/transformed Perfect Character that acts and should direct our technology. In this spiritually aware orientation *to* the Greatest Good we can use our scientific knowledge *for* the Greatest Good.

Mirrors

To discuss activity of souls, Drummond used one of the most classic images of spirituality and psycho-sociality from Plato to Lacan: the mirror. Reiterating his conviction that spirituality works by principles of natural laws, Drummond stated that one of the first laws of nature is that we all "are mirrors" and "all human intercourse is a seeing of reflections."[63] Like Origen and Augustine, Drummond maintained that "Whether we like it or not, we live in glass houses. The mind, the memory, the soul, is simply a vast chamber panelled with looking-glass." We receive and reflect impressions, but we are not completely passive.[64]

Drummond insisted that the divine—or ideal—view of the physical world does not ignore its reality. In "Clairvoyance" Drummond wrote that "This earth is not merely a place to live in, but *to see* in. We are to pass through it as clairvoyants, holding the whole temporal world as a vast

[62] See David Grumett, "Transformation and the End of Enhancement: Insights from Pierre Teilhard de Chardin" in Cole-Turner, *Transhumanism and Transcendence*, 37–49; Tucker, "The Singularity and Human Destiny," 1–11.

[63] Drummond, "The Changed Life," in Maine, *The Greatest Thing in the World by Henry Drummond*, 103.

[64] Ibid.

transparency, through which the eternal shines."[65] Such spiritual sight views materials as more than their "market value." Although some transhumanists or religious ascetics might ignore the physical world in pursuit of spiritual self-fetishization, Drummond believed that we "should look at the things which are seen," including our fellow beings, with moral vision and imagination.[66] It is through "these homely temporals"—our material culture and challenges of everyday life—that we "learn" the divine perspective of Love.[67] Drummond cited Browning's articulation:

> Earth's crammed with Heaven,
> And every common bush afire with God:
> But only he who sees, takes off his shoes.[68]

According to natural laws, the self's mirror cannot exist in isolation. Our will, the self's existential "I" is involved one way or another with the rest of creation. Just as the will must become aware of its existence and reflection in the divine mirror, it must become aware of others' reflections as well. In this way, we are each "an arrangement of mirrors" who meets other people's arrangements of mirrors every day.[69] Drummond used the term "involution"[70] to describe this mutual influence human beings have on each other, similar to what existentialists call the "intersubjectivity" or "interconnectedness" of every decision we make and every action we take.[71] In addition to our resemblances explained by laws of physical reproduction, we develop, Drummond said, spiritual resemblances in our "outlook" on life. Our "soul chambers" are filled with parts of others' souls[72]:

[65] Drummond, "Clairvoyance" in Maine, *The Greatest Thing in the World by Henry Drummond*, 185.

[66] Ibid., 182–189.

[67] Ibid.

[68] Ibid., 191 citing Browning in *Aurora Leigh*.

[69] Drummond, "The Changed Life" in Maine, *The Greatest Thing in the World by Henry Drummond*, 103–4.

[70] *Ascent of Man*, 30–31; Thomson, *Evolution and Involution*.

[71] Sartre, *Existentialism and Human Emotions*; Celia Deane-Drummond, "Taking Leave of the Animal? The Theological and Ethical Implications of Transhuman Projects" in Cole-Turner, *Transhumanism and Transcendence*, 122–123; Stephen Garner, "The Hopeful Cyborg" in Cole-Turner, *Transhumanism and Transcendence*, 97.

[72] Drummond, "The Changed Life" in Maine, *The Greatest Thing in the World by Henry Drummond*, 105.

> These things, these books, these events, these influences … are life and death, beauty and deformity. When once the image of likeness of any of these is fairly presented to the soul, no power on earth can hinder two things happening—it must be absorbed into the soul and forever reflected back again from character.[73]

In other words, what goes around eventually comes back around. Drummond used as illustrations the Biblical stories of David and Jonathan,[74] Victor Hugo's characters in *Les Misérables*, and examples of couples said to look alike after living together for many years. He even criticized John Bunyan's Pilgrim for being self-absorbed and concerned only about his own arrival at the Celestial City; the more mature Pilgrim would contemplate the power of involution in spiritual progress.[75] This "making of higher and better" human beings, Drummond averred, occurs not through self-techniques but through mutually interactive relationships with God and other humans.

Drummond's mirror analogy anticipated modern psycho-social neurobiology and theories of *simulation processes.* Neuroscientists today, for example, study how individuals "identify similar processing involved in both the *experience* of an intention or emotion in oneself and the *perception* of an intention or emotion in another."[76] This becomes even more complex with our technologically integrated lives. As Michael Spezio explains, "The networks carrying this double duty of self-representation and other-representation are termed 'shared circuits' or sometimes 'mirror neurons.'"[77] This can allow us to "see 'ourselves as we really are.'"[78] And it raises important questions about individuality in a rapidly transhumanizing society.[79] Our own spiritual improvement cannot be isolated from that of others. Every encounter, every event, every conversation, every social media post involves our own spiritual progress *and* that of others.[80]

[73] Ibid., 104.

[74] David, for example, lived with Saul and Jonathan in Saul's household and then eventually took Jonathan's son into his own household after Jonathan's death.

[75] Drummond, "The Programme of Christianity" in Maine, *The Greatest Thing in the World by Henry Drummond.*

[76] Spezio, Michael, "Human or Vulcan? Theological Consideration of Emotional Control Enhancement" in Cole-Turner, *Transhumanism and Transcendence*, 156–7.

[77] Ibid., 157.

[78] Ibid., 157.

[79] Grumett, David, "Transformation and the End of Enhancement: Insights from Pierre Teilhard de Chardin" in Cole-Turner, *Transhumanism and Transcendence*, 37–47.

[80] Drummond, *The Greatest Thing in the World*, Samford edition, 36.

Because of the effects of such soul-mirroring, Drummond wrote that we should strive to be nothing less than "sanctifiers of souls" and recognize others who are.[81] Our mirrors should reflect "unselfishness, sympathy, and self-sacrifice for Others."[82] Echoing the Duke of Argyll, Drummond called this altruism a "new and very affected name for the old familiar things which we used to call Charity, Philanthropy, and Love."[83] In Augustinian terms: "Of all those who are capable of enjoying God together with us, we love some whom we are helping, and some who are helping us; some whose help we need and some whose needs we are meeting; some to whom we give no benefit and some by whom we do not expect any benefit to be given to us. But it should be our desire that they all love God together with us, and all the help that we give to or receive from them must be related to this one end."[84]

Do our choices improve our selves in God's eyes? Do they benefit or diminish Love in others' humanity? Are our acts those of altruism or self-fetishization? Our failure to recognize and value the existence, experiences, and struggles of others as much as ourselves—in our natural resources, digital communications, material culture, and virtual reality—cannot destroy the energy of Love, but it can slow or obscure it temporarily, which is mirrored back to us. If, for example, our efforts to improve our own selves, our own conditions, and our own fortunes diminish the humanity of others, our will is not acting fully in Love, and we are not implementing the Greatest Good. This failure to implement the Greatest Good and reflect the Perfect Character of Love allows and perpetuates suffering. Drummond believed that human misery is overcome by the omnipotent energy of Love flowing through the Perfect Character in our decisions and actions.

Advolution

Instead of "besting" other individuals with our scientific knowledge and technological innovations, Drummond's model shifts focus to ways we participate in the proliferation of Love by using our scientific and technological advancements in the Character's "Struggle for Others."[85]

[81] Drummond, "The Changed Life," in Maine, *The Greatest Thing in the World by Henry Drummond*, 104.

[82] Drummond, *Ascent of Man*, 30–31.

[83] Ibid., 345, quoting The Duke of Argyll, *Edinburgh Review*, April, 1894.

[84] Augustine, *On Christian Teaching*, 21–22.

[85] Drummond, *Ascent of Man*, 346.

Drummond imagined a "still wider evolution" in which the formation of "Christ Consciousness" gives human beings "the power to create a new unity based on divine grace."[86] As with the law of mirrors, our attention to Love in others' lives does not just reproduce equal amounts; it multiplies exponentially. Drummond called this proliferation of Love "advolution," a rolling momentum greater than the total individual efforts.[87] Advolution describes a process by which exponential proliferation occurs in such small steps that it is often imperceptible except in hindsight or to those intentionally looking for it.

To allow the Perfect Character to flow through us and others, gathering strength as it is exercised and shared, Drummond proposed that "each shall explore with new respect the other's world, and, instead of delighting to accentuate their contrasts, strive to magnify their infinite harmonies."[88] Everyday challenges by other people—family, friends, colleagues, strangers, and enemies—is a school of opportunities to learn more of the Perfect Character. The Character develops not in solitude, but in the "stream of life" and proliferates accordingly.[89] "Life," Drummond echoed Goethe, "is … an education," and life's greatest lesson "is *how better can we love.*"[90] Applying Drummond's philosophy to transhumanism, we examine how we are embodying and expressing the Perfect Character in our interactions with other beings—physically, digitally, and every other way. The ability to "see" and "be" the qualities of Love requires regular attention to the inmost mirror.

The Practice

One of Drummond's greatest gifts to modern and postmodern audiences is a simple ontological exercise to help us shift our self-absorbed focus on our own existence to God's "larger view" of ourselves and others.[91] Drummond called the daily discipline a "mirror setting" of our human will to the Perfect Character. It can be practiced by anyone and is for everyone, including those in a rapidly changing technological society.

[86] George, Timothy, "Forward" in *The Greatest Thing in the World*, Samford edition, viii.
[87] Drummond, *Natural Law in the Spiritual World*, 390.
[88] Drummond, *Ascent of Man,* 321.
[89] Drummond, *The Greatest Thing in the World*, Samford edition, 51.
[90] Ibid., 49.
[91] Drummond, *Natural Law in the Spiritual World*, 376; 1 Cor. 13: 4.

Drummond theorized that we each need to pause for internal reflection every 24 hours in order to choose the "angle" or "program" from which we will automatically act and react for the rest of the day.[92] Habitual exercising and training the human will by recognizing the divine will is crucial. After critiquing popular devotional methods, Drummond advocated a two-to-ten-minute daily pause to reflect intentionally on our interactions with God and others. Such an exercise will "turn [your mirror] to [God]" and "set [it] at the right angle" to keep the surface of your own mirror bright, and ever in position to "uncover the face which is to look at Christ."[93] Re-orienting our focus from *chronos* (calendar, clock) time to *kairos* (eternal) time helps us see the "larger view" of life and cultivate the first quality of Perfect Love: patience.[94] We need this, Drummond wrote, because our natural world moves and changes every moment of every day. Therefore, "each day, each hour, demands a further motion and re-adjustment of the soul."[95] Such daily internal "re-adjustment" will, Drummond speculated, "move the vast inertia of the soul, and place it, and keep it, where the spiritual forces will act upon it."[96] According to Drummond, the spiritual discipline works by natural laws of gradual progress and general advolution.[97] In practical terms, he equated the cleaning and "adjusting" of our spiritual mirror to physical grooming before we go out and meet others. During this practice, we meet "face to face" the One in Whom we "live and move and have [our] being."[98]

Drummond advocated spending this two-minute pause in *lectio divina*—reading Scripture aloud. And he specifically recommended reading 1 Corinthians 13 every day for three weeks. (It takes most humans about two minutes to read the entire passage.) Many theologians and philosophers since the first century CE have considered this "Hymn to Love" the greatest text for all humanity, for all time. Like Drummond, we can apply

[92] Drummond, *Natural Law in the Spiritual World, 286.*

[93] Drummond, "The Changed Life" in Maine, *The Greatest Thing in the World by Henry Drummond*, 115.

[94] Drummond, *Natural Law in the Spiritual World*, 376; 1 Corinthians 13:4.

[95] Drummond, "Changed Life" in Maine, *The Greatest Thing in the World by Henry Drummond*, 115; Mark 12:28–31; Matthew 22:34–40; Luke 10:25–37.

[96] Drummond, "The Changed Life" in Maine, *The Greatest Thing in the World by Henry Drummond*, 115.

[97] Drummond, *The Greatest Thing in the World*, Samford edition, 49.

[98] Acts 17:28; Drummond, *The Greatest Thing in the World*; Drummond, *Natural Law in the Spiritual World*, 259.

ourselves to multi-disciplinary studies of its history and Greco-Roman-Jewish influences. We can engage its etymology and hermeneutical possibilities; read its uses by Neo-Platonists, Scholastics, and Mystics; and find its cross-cultural applications. Most importantly, however, we can *practice* it. Drummond's thesis is that we participate in the Greatest Good every time we "remember" 1 Corinthians 13. As we speak it, we remember "That Which Is Perfect" in our inmost being, or soul memory, and our souls re-connect with the Communion of Saints in *kairos* time. The more frequently we remember the Perfect Character, the further into our "memory halls" and consciousness it goes. This is why Augustine began Book X of *Confessions,* about human memory, with 1 Corinthians 13.[99]

Our souls "learn love" by the exercise of the Perfect Character in our interactions with others. Regularly focusing our attention, voices, and memories on a Scripture passage that transcends cultural and chronological boundaries habituates our will. It reminds us what "the most excellent way" is.[100] It prepares us to see more clearly and realistically our daily "schoolroom" of each decision, activity, and encounter—including technological ones—according to "That Which Is Perfect."[101] Knowing this passage "by heart" helps us approach all of life's challenging questions, difficult decisions, contentious conversations, and soul-wrenching events. Instead of losing our divinely created human will, we find our will transformed into the more Perfect Character.[102] For Drummond, as our will grows closer to That Which Is Perfect, our humanity—the *Imago Dei*—is not lost but more fully found as God created us to be. Our will becomes an "organic reflection" of That Which Is Perfect, "naturally" embodying rather than artificially mimicking the qualities personified in 1 Corinthians 13. From inside That Which Is Perfect, we determine the actions we take in the physical world. We express the transcendent law of Love and radiate Love of God and others in all directions.[103]

[99] Augustine, *Confessions*, Book X.

[100] 1 Corinthians 12:31.

[101] Drummond, "The Programme of Christianity" in Maine, *The Greatest Thing in the World by Henry Drummond*; Drummond, "The Changed Life" in Maine, *The Greatest Thing in the World by Henry Drummond*, 98; Drummond, *Natural Law in the Spiritual World*, 286, 390; Drummond, *The Greatest Thing in the World.*

[102] Drummond, "The Changed Life," in Maine, *The Greatest Thing in the World by Henry Drummond*, 98.

[103] Drummond, *The Greatest Thing in the World*, Samford edition, 36.

Noting many people's impatience and desire for convenience in spiritual growth,[104] Drummond contended that our soul does more than merely "imitate" Christ "mechanically" and that spiritual maturity does not occur overnight.[105] Drummond provided a textual image of how he believed such gradual transformation happens. He arranged 2 Corinthians 3:18 as a chiasmus, or textual mirror, like 1 Corinthians 13. He then paraphrased the verse, interchanging Character and Glory.[106]

> We all
> With unveiled face
> Reflecting
> As a mirror
> The Glory [Character] of the Lord
> Are transformed
> Into the same image
> From Glory [Character] to Glory [Character]
> Even as from the Lord
> The Spirit.

Conclusion

The spiritual and physical habits that we cultivate and allow, including how we use our scientific and technological innovations, matter to humanity and the rest of Creation. As the rate of our technological innovations speeds up, we are challenged to exercise our human will in the disciplines of Love. We find ourselves in brand new "schoolrooms" every day regarding humanity and humaneness. It is easy for us to allow our technological culture to determine our mental, physical, and spiritual habits rather than the other way around. We can, however, use Drummond's devotional mirror to examine our motives and methods on a daily basis and not just on special occasions of extreme jubilation, stress, or desperation. As we make decisions about medical technology, drug options, artificial intelligence, and social media, for example, we can act less out of self-fetishization, envy, pride, impatience, stinginess/hoarding, rudeness, fear, isolation, insincerity, unkindness, and dull passivity in our technological cultural milieu.

[104] Drummond, "The Changed Life," in Maine, *The Greatest Thing in the World by Henry Drummond*, 98.

[105] Ibid., 103.

[106] Ibid., 98.

Drummond's spiritual practice remains useful and effective for such questions in our personal, professional, and public/political lives today. The nine terms Drummond used to enumerate the qualities of Perfect Character of 1 Corinthians 13—Patience, Kindness, Generosity, Humility, Courtesy, Unselfishness, Good Temper, Guilelessness, and Sincerity—may seem unusually challenging for our technological society, which is an indication that we need diligent attention to them. Drummond called them everyday "opportunities" to learn and participate in Love. When we exercise these nine interactive ingredients, we activate the Greatest Good by "becoming" and "being" Beatitudes (Blessings, Goodness).[107] What is *Generous Patience* in a culture obsessed with speed and convenience? What is *Courteous Kindness* in an encounter with a self-absorbed individual? What is *Humble Unselfishness* with technology that gives me advantages over others? Am I *Guileless* and *Sincere* in all of my communications?

I believe that Drummond was correct in his assessment of our need to bring the full Love of God and the Golden Rule into every part of our lives, including how we choose to live with our science and technology. When we are oriented to the Perfect Character of Love we use our science and technology for the Greatest Good. Toward that end, the Church can promote Drummond's practical spiritual discipline individually and communally. This attention and orientation to Love can transform self-centered, isolating, and potentially destructive struggles for existence into more meaningful, transcendent living.[108] As Drummond exhorted, "life's one charge" is the formation of the divine will to Love in us.[109] Everything else revolves around it.[110]

> 'Where Love is, God is. He that dwelleth in Love dwelleth in God. God is Love.' Therefore love. Without distinction, without calculation, without procrastination, love. Lavish it upon the poor ... the rich.... our equals....[111]

[107] The Sermon on the Mount, Matthew 5–7.

[108] Drummond, *The Greatest Thing in the World*, Samford edition, 38.

[109] Drummond, "The Changed, Life," in Maine, *The Greatest Thing in the World by Henry Drummond*, 116.

[110] Ibid.

[111] Drummond, *The Greatest Thing in the World*, Samford edition 61.

Bibliography

Augustine. 1991. *Confessions: Book X*. Translated and Introduction by Henry Chadwick. Oxford: Oxford University Press.

———. 2008. *On Christian Teaching*. Translated and Introduction by R. P. H. Green. Oxford: Oxford University Press.

Bimber, Bruce. 1994. Three Faces of Technological Determinism. In *Does Technology Drive History? The Dilemma of Technological Determinism*, ed. Merritt Roe Smith and Leo Marx. Cambridge, MA: MIT Press.

Bowen, Francis. 1881. Christian Metempsychosis. *The Princeton Review* 1: 315–341.

Browning, Robert. 1864a. *Dramatis Personae*. London: Chapman and Hall.

Browning, Elizabeth Barrett. 1864b. *Aurora Leigh*. London: J. Miller. Accessible at http://digital.library.upenn.edu/women/barrett/aurora/aurora.html

Browning, Robert. 2004. *Saul*. Whitefish: Kessinger.

Cole-Turner, Ronald, ed. 2011. *Transhumanism and Transcendence: Christian Hope in an Age of Technological Development*. Washington, DC: Georgetown University Press.

Corts, Thomas. E., ed. 1999. *Henry Drummond: A Perpetual Benediction*. Edinburgh: T & T Clark.

Corts, Thomas E., and Marla Haas Corts. 2003. The Man Who 'Lived in First Corinthians 13'. In *The Greatest Thing in the World by Henry Drummond*, 1–25. Birmingham: Samford University Press.

Darwin, Charles. 1982. *The Origin of Species*. New York: Penguin.

Drummond, Henry. 1894. Christianity and Evolution. In *The World's Congress of Religions at the World's Columbian Exposition: The Addresses and Papers Delivered Before the Parliament, August 25–October 15, 1893*, ed. John Wesley Hanson. Chicago: International Publishing.

———. 2003a. In *The Greatest Thing in the World*, ed. Sandra L. O'Brien. Birmingham: Sanford University Press.

———. 2003b. *Natural Law in the Spiritual World*. Whitefish: Kessinger.

———. 2007. *The Ascent of Man*. New York: Cosimo.

Emilsen, William W. 2001. Gandhi, Scripture, and the Bible. *Pacifica* 14: 83–84.

Ewen, Stuart. 1976. *Captains of Consciousness: Advertising and the Social Roots of the Consumer Culture*. New York: McGraw-Hill.

Hardesty, Nancy. 2003. *Faith Cure: Divine Healing in the Holiness and Pentecostal Movements*. Grand Rapids: Baker.

Kern, Stephen. 2003. *The Culture of Time and Space, 1880–1918*. Cambridge, MA: Harvard University Press.

King, Ursula. 2004. Love—A Higher Form of Human Energy in the Work of Teilhard de Chardin and Sorokin. *Zygon* 39 (1): 77–102.

Kranzberg, Melvin. 1986. Technology and History: 'Kranzberg's Laws'. *Technology and Culture* 27 (3): 544–560.

Lightman, Bernard. 2010. Darwin and the Popularization of Evolution. *Notes and Records of the Royal Society* 64: 17–19.

Linnaeus, Carl. 2005. *Philosophia Botanica*. Trans. Stephen Freer. New York: Oxford University Press.

Maine, G.F., ed. 1953. *The Greatest Thing in the World*. London: Collins.

Marvin, Carolyn. 1990. *When Old Technologies Were New: Thinking About Electric Communication in the Late Nineteenth Century*. Oxford: Oxford University Press.

McGinn, Bernard, ed. 2006. *The Essential Writings of Christian Mysticism*. New York: Modern Library.

More, Max. 1993. *Technological Self-Transformation*. www.Maxmore.com/self-trns.htm

Richard, Timothy. 1910. *The New Testament of Higher Buddhism*. Edinburgh: T & T Clark.

Roughley, Neil, ed. 2000. *Being Humans: Anthropological Universality and Particularity in Transdisciplinary Perspectives*. New York: Walter de Gruyter.

Sartre, Jean-Paul. 1957. *Existentialism and Human Emotions*. New York: Philosophical Library.

Scott, Anne. 2004. 'Visible Incarnations of the Unseen': Henry Drummond and the Practice of Typological Exegesis. *The British Journal for the History of Science* 37 (4): 435–454.

Smith, George Adam. 1901. *The Life of Henry Drummond*. London: Hodder & Stoughton.

Sorokin, Pitirim A. 2002. *The Ways and Power of Love: Types, Factors, and Techniques of Moral Transformation*. Radnor: Templeton Foundation Press.

Thomason, George. 1880. *Evolution and Involution*. London: Trübner and Co.

Transmigration of Souls. *The Columbia Encyclopedia, 6th edition*. New York: Columbia University Press, February 15, 2016. www.encyclopedia.com

Tucker, Patrick. 2006. The Singularity and Human Destiny. *The Futurist*, March–April, 1–11. http://www.singularity.com/KurzweilFuturist.pdf

CHAPTER 5

Christianity's Rigged Debate with Transhumanism

Boaz Goss

Introduction

The first Christians living in the ancient Roman Empire knew how to raise hell. I mean that quite literally, since the pagans who upheld the traditional religious status quo looked upon the Christian philosophy and saw "the very order of the cosmos collapsing at its base, drawing everything down into the general ruin and obscene squalor of a common humanity.

This chapter is dedicated to my father, Tom Bradley Zechariah Goss, who drove four hours to hear it presented at the original conference but died before seeing it published. This chapter is also dedicated to two of my best friends, Angela Lange and Weston Koenn, who have brought Christ into my life through their co-suffering with me. To receive so much love in a year of so much heartache is to have a dream come true in the midst of a nightmare and redeem it.

B. Goss (✉)
Albert Gnaegi Center for Health Care Ethics, Saint Louis University,
St. Louis, MO, USA
e-mail: boaz.goss@slu.edu

S. Donaldson, R. Cole-Turner (eds.), *Christian Perspectives on Transhumanism and the Church*, Palgrave Studies in the Future of Humanity and its Successors,
https://doi.org/10.1007/978-3-319-90323-1_5

A spectacle of monstrous impiety and wicked degeneracy."[1] The Christian belief in the God-Man Christ led them to attack the pagan Roman social institutions of sex slavery, abortion, and caste hierarchy, institutions which had been established by the gods themselves and assumed to be eternal features of an endlessly repeating universe.[2] By disrupting the traditional pagan lifestyle, Christians threatened to upset the traditional rhythm of sacrifices. The pagan gods needed to feed off the efforts of humanity in exchange for holding back the dismal tide of chaos, so starving pagan society of the human sacrifices it needed to function meant starving the gods; meant unleashing the tides of hell upon the universe. Thus the standard accusations against the Christians were "anarchists" and "atheists," and these were not misnomers.[3] Romans treated the first Christians the same way that Americans treat communists: ultra-progressive, utopian atheists attacking our "good old religion and way of life." But this seems entirely wrong, given that ever since the Enlightenment up until the present-day exchange with transhumanism, Christianity has been portrayed as an enforcer of the status quo, a method of socio-political control, and an enslaving tyrant of human ingenuity.

If we turn to present-day Chinese Christianity—flourishing despite persecution from the communist government—we hear a similarly jarring story: Chinese intellectuals are flocking to Christianity. The Chinese youth view Christianity as a progressive rival to traditional Chinese social religions: Confucianism, Buddhism and Taoism are seen as Luddite relics, while Christianity provides a pathway into industry and technology.[4] "Luddite" is an insult some transhumanists lob against Christianity, since Christianity supposedly considers technological progress to be blasphemy against God's sovereignty. Yet right now on the other side of the world the opposite story is spreading: Christianity is a beacon of progress, an incubator of innovation against dead-end traditions. Modern Chinese Christianity strikingly resembles the first few generations of Christians—the "forerunners of the faith" as they are known in Catholic and Eastern Orthodox circles—while whatever kind of Christianity that transhumanists are attacking sounds much closer to the nationalistic pagan religion of Rome.

[1] Hart, *Atheist Delusions*, 169.
[2] Ibid., 201.
[3] Ferngren, *Medicine and Religion*, 48.
[4] Stark, *A Star in the East.*

How did modern Christianity come to be portrayed as its ancient pagan arch enemy? I argue that through a series of political and philosophical shifts foundational to our current age, Christianity has been torn to pieces and lost a propaganda war. These shifts caused Christianity to lose the intimate bond it previously held with what we now call the "scientific" disciplines. A gulf between "religion" and "science" was created, and the link between the two was no longer clear. In fact religion and science appeared to be fighting. At the same time, endeavors like "ethics" were placed in the realm of "religion," while endeavors like "technology and practical knowledge" were placed in the realm of "science." With modern life framed in such a way, Christianity was left as an idle bystander to scientific pursuits, its only function being able to make ethical objections to science's progressive march. In the same way that contemporary bioethicists are viewed by doctors as doing nothing but inhibiting vital scientific progress, Christianity will naturally be demonized by those carrying the banner of scientific progress. Thus I argue that Christianity's debate with transhumanism was rigged before it even began. I end by detailing how Christian reactions to evolution lie at the heart of this problem, and offer some suggestions—including some fundamental theological overhauls—for stitching Christianity, the natural sciences, and technology back together.

Christianity Before There Was Religion

We must first realize that our modern tendency to separate the theological realm from the political, scientific, and social is a very new arrangement. Before the Reformation, all known societies used the picture of the universe painted by their priests to structure every aspect of their civilizations. Cultures of every continent accepted that the universe played a moral harmony that all socio-political institutions had to match in order to thrive. There was no separation between religion and state; the state structured itself around religion, as did every other form of human culture. The state attempted to preserve and enforce the divine laws. Family and friends were obliged to help each other fulfill their God-given destiny. Before the Reformation, the physical sciences were studies of the universe that was a reflection of God's nature, and thus scientific discoveries were expected to lead the researcher into spiritual contemplation, since their object of study pointed toward God.

In this socio-political framework there were no separate spaces for religion and science. Only a few in classical and medieval history were even able to conceive of a secular realm, and human activity had the ultimate goal of reflecting the divine. Even the pagan Greeks and Romans understood this; they merely differed with Christians over which gods were the role models. Plato's Republic, modeled after the divine Forms and Antigone's martyrdom for obeying the laws of the gods in spite of the laws of king Creon, exemplify this framework that would eventually be Christianized.

Because the political, the scientific, and the social were bundled together, the sphere of culture we now call "religion" did not exist. The sphere of culture we now call "science" did not exist. Our current attitudes and ideas about both depend on them being isolated into their own separate secularized realms. The Latin words that we derive these two terms from did exist in ancient times but they signified good personal moral and intellectual habits, along with the practices needed to exercise those habits.[5]

"Religion" was the moral virtue of committed trust. Before the Reformation it was most commonly known by its Latin form, *religio*. If we were to ask someone from Classical-Medieval times about their religion, they would tell us about the habits they practice to serve and preserve their dearest relationships: their family, their church, their town, their god.[6] Christianity and the other worldviews we now call "religions" were not called this until the political upheavals of the Reformation in the early and middle 1500s. Thus as strange as it may sound to our modern ears, Christianity was not a religion, but it contained many religions.

This makes more sense if we realize that "religion" was defined like a "vow." There are many kinds of vows and each comes with different habits needed to uphold these vows, but one can make and uphold many vows without any of them necessarily conflicting. We can make certain vows to our family, to our spouse, to our church, even to God Himself, and this does not necessarily lead to an immoral or incoherent life. Christianity itself was not equivalent to these vows, but was a worldview that wove them together into a grand picture, providing them with moral guidance and transcendent purpose. This grand picture had to be drawn and maintained by theology, providing a critical foundation for Christianity. While religion and theology were thus closely linked, there was a major conceptual difference: theology was a science.

[5] Harrison, *The Territories of Science and Religion*, Chapter 1.

[6] Ibid., 8. See also "The Invention of Religion" in Cavanaugh, *The Myth of Religious Violence*.

Defining theology as a science will likely strike our modern ears as an oxymoron, since today we reserve the term "science" specifically for the natural sciences: fields of empirical study like biology, physics, and chemistry devoted to purely physical objects.[7] Yet up until about the mid-1850s, any discipline that employed logic to discover knowledge was called a *scientia*, the Latin word for "science."[8] For this reason the liberal arts were originally known as the "liberal sciences."

Those who practiced the scientific disciplines would develop three intellectual virtues: understanding (grasping first principles and axioms), science (using deductive and inductive logic on top of the first principles and axioms to discover new truths), and wisdom (letting these discoveries lead to deeper contemplation of reality and God).[9] Thus science had two distinct but related definitions: a science was a comprehensive field of study, but also the virtue of logical skill.

In this Classical-Medieval understanding, the relationship between religion, science, and their distinct practices was clear and mutually beneficial. The Christian philosophy provided a worldview that made the whole universe intelligible.[10] Because Christianity insisted from the beginning that the divine Logos had birthed everything that exists, and that this creation was intricately harmonic, rational, and beautiful throughout, this story would very naturally guide believers toward study of the world around them, in order to catch a glimpse of the Logos orchestrating all things.[11] The scientific ethos within Christianity thus considered the purpose of scientific inquiry not simply practical knowledge, but moral character formation.[12] By intently observing a creation held in existence by God Himself, something divine might be revealed. Christians would learn the inner order of the world, which would also teach them something about humanity. They could use this knowledge to create more harmony in their dearest relationships, fulfilling their "religious" vows. In this way, the intellectual and moral realms were so interwoven that they were both considered essentials to the good life.

[7] Harrison, *The Territories of Science and Religion*, Chapter 6, section "Science and its Methods," 164–170.

[8] Harrison, *The Territories of Science and Religion*, 13–14.

[9] Ibid., 12.

[10] Hart, "Divine Humanity" in *Atheist Delusions*.

[11] Pope Benedict XVI, "Faith, Reason, and the University: Regensburg Address" (University of Regensburg, 2006).

[12] Harrison, *The Territories of Science and Religion*, 33.

The "religion" of Christianity was primarily concerned with relationships, not with ideas. Theology, as the "science" of Christianity, was primarily responsible for engaging the intellectual realm of ideas. This does not mean that theology and its doctrines were unimportant in the Christian life. It was critical to loyally devote oneself to the proper relationships, but also necessary to grasp the truth with one's mind. In the present day, certain Christians advocate that there is a division between theology and relationships, sometimes known as the conflict between "doctrine versus pastoral practice." Christians who accept this division typically conclude that relationships are more important, and that doctrine deserves less attention by Christians or that doctrine must change in response to relationships. It might appear that by having a distinct virtue for relationships (*religio*) and a distinct virtue for knowledge (*scientia*), Classical-Medieval Christianity is likewise implying that there is a division between relationships and doctrine. This would be a serious misunderstanding. Remember that Christianity itself was not considered a religion, rather religion was only a single virtue in the life of a Christian. A proper Christian life consisted of many other virtues—not simply the virtues concerning good relationships—and all virtues were interdependent on the others. The Christian ideal was a smart saint—someone who excelled in both the virtue of *religio* by being faithful in their relationships and the virtue of *scientia* by being intelligent. The religious and the scientific virtues both had effects on one's moral character, consequently it was important that Christians grew in both. Failing to develop either virtue meant that a Christian was missing an important part of their God-given self. Thus the debate today over whether Christians ought to focus on either doctrine or relationships would make little sense to Christians in past eras, since they would claim both are necessary for the fullness of Christian life and both complement the other without conflict. As the Eastern Catholic-Orthodox maxim goes, "orthodoxy requires orthopraxy." One must live a holy life (orthopraxy) to discover truth, and one must discover truth (orthodoxy) to live a holy life.

The War Against Christianity That the Reformation Started

The symbiotic unity between religion and science was lost following the Reformation. It is at this point that "religion" and "science" start to take on the definitions they still have today, along with their conflicted relationship.

The splintering of Christianity caused riots all across Western Europe between Protestants and Catholics. Both factions understood that the ancient link between theology and the rest of human culture meant that a different theology would lead to a different political order, thus we can understand why both factions attacked each other as political traitors and not simply heretics. The entire structure of the universe was at stake, not simply private ideas about heaven. The stakes were high and political—it seemed as if all of Europe might balkanize into an ever-warring mass of Protestants against Catholics.

A couple of treaties across the continent require our attention here. In England there was Queen Elizabeth I's Act of Supremacy in 1558, which required all Catholics and Protestants holding public office to swear their primary allegiance to the state, thereby creating the first modern secular realm. Since both Catholics and Puritans were causing politically threatening riots, they had to be kept outside of state matters. But Elizabeth was glad to tolerate philosophical differences as long as they were not offensive enough to incite violence or treason, an attitude strikingly familiar to modern attitudes toward religion.

A few years before, the Augsburg Peace Treaty of 1555 settled tension between Catholic Charles V of the Holy Roman Empire and the Schmalkaldic League of Lutheran Royalty. It was agreed that land would be divided according to the beliefs of the inhabitants ("whose land, his religion").[13] But this required having litmus tests to prove who believed what. Thus the treaty specified that there were two "religions," Catholicism and Lutheranism, and they were defined by their confessions and creeds. This created a subtle but monumental redefinition of "religion"—one's religion was no longer the habits of loyal commitment to a church community, but rather a set of propositions that one believed to be true.[14] The essence of religion was no longer found in the hearts and minds of particular church communities, but in a set of answers written in books. And apparently these answers made people violent.

[13] Ibid., 97.

[14] Ibid., Chapter 4.

Science Discovers "God-of-the-Gaps" Doesn't Exist

These answers were not only violent, but demonstrably wrong, as scientists increasingly came to believe. Before the Reformation era, scientists were called "natural philosophers" because their research led to philosophical reflection as well as empirical data.[15] Today scientists only ask "how" questions, but in earlier times they also explored "why" questions. Galileo was a fervent opponent of "why" questions—to him they were uninteresting and uninformative.[16] He cared about getting pragmatic answers that would benefit society through technological advancement, and his successes inspired later generations to not only adopt his theories, but also his method and attitude.[17] Asking "why" questions which required talking about immaterial objects—including God—became a sign of ignorance to Enlightenment era scientists, even though these scientists were still mostly Christians. The goal of science was to find answers to the only real questions—the "how" questions—and do so without discussing philosophical or immaterial concepts.

Traditional Catholic-Orthodox doctrine held that God was the answer to the "why" questions of the universe. There was thus plenty of room for a god and one could propose a purely naturalistic explanation of how things fall (gravity) without that infringing upon the question of why things even exist to fall in the first place (God's love). But "why" questions were no longer valid questions according to scientists who inherited Galileo's attitude. Consequently Western intellectuals were brought to a dilemma: either God was the direct physical explanation of gravity, or magnetism,[18] or the regular orbit of planets,[19] or God was not involved at all. And Western scientists for over two centuries were perpetually realizing that a god was not actually in the unexplained gaps between our scientific knowledge. It looked from this perspective that religion had posited an absurd entity that did not exist to explain motions that could be explained by matter and mathematics alone. As astronomer Pierre-Simon Laplace said in 1802 to Napoleon, "God? I have no need of that hypothesis."[20]

[15] Ibid., 26–34.
[16] Burtt, *The Metaphysical Foundations of Modern Physical Science*, Chapter 3: Galileo.
[17] Ibid., 33–35.
[18] Ibid., 198.
[19] Ibid., 287–302.
[20] Barbour, *Religion & Science*, 35.

Christianity: The Bioethicist Religion

Framed like this, Christianity becomes a private set of beliefs concerning all the things that science discarded as either "uninteresting" or "non-existent." This includes not only God, but also philosophical entities like ethics. Christianity in the modern age is supposed to limit itself to concerns about ethics and other imaginary notions but, at the same time, Christians are not supposed to disrupt society with their ethics. Christianity is given a realm that Western seculars consider non-existent, and also irrelevant. The attitude is captured well by Stephen Hawking in his recent book *The Grand Design*: "Philosophy is dead. Scientists have become the bearers of the torch of discovery in our quest for knowledge."[21]

This puts Christianity on the wrong side of the power relationship. Apparently the Christian's role is to believe in immaterial objects that do not exist, or do not matter, or both. Meanwhile science is free to benefit humanity with increasing efficiency. Science has all the tools and technical knowledge, Christianity has imaginary ideals that can at best cheerlead scientists into doing what they already wanted to do anyways. At worst, Christianity lives up to Nietzsche's "life-negating slave morality" label by attempting to halt scientific progress with "ethics" and other unscientific notions.

Christianity's relationship to transhumanism thus parallels the relationship between bioethics and medicine. Christianity acts as the bioethicist, doing external ethical oversight, while transhumanism acts as the doctor, claiming all the technology and knowledge to benefit humanity.

Judging from the history of medicine, this is a precarious position because the American medical establishment has consistently attempted to bypass external ethical oversight. Medical history makes this point very clear—scientists do not like outsiders trying to police them. Many doctors find bioethicists to be bothersome nuisances that need to be marginalized as much as possible. We see this attitude in a recent op-ed by Harvard psychologist Steven Pinker:

> Biomedical research promises vast increases in life, health, and flourishing. Given this potential bonanza, the primary moral goal for today's bioethics can be summarized in a single sentence. Get out of the way. A truly ethical bioethicist should not bog down research in red tape, moratoria, or threats of prosecution based on nebulous but sweeping principles such as "dignity," "sacredness," or "social justice."[22]

[21] Hawking and Mlodinow, *The Grand Design*, 5.
[22] Pinker, "The Moral Imperative for Bioethics."

Pinker's attitude forces the ethical overseer into an unwinnable dilemma. The first option is that she can condemn what the scientist wants to do, in which case she will be villainized as a Luddite obstructing progress and the scientist will begin searching for a way to bypass the ethicist altogether. As for the second option, the bioethicist can approve of the scientific project, but then the bioethicist becomes superfluous, being untrained in the technical skills needed to carry out the project.

The transhumanist attitude toward Christianity mirrors Pinker's attitude toward bioethicists. Christianity is the inhibitor of technological progress, a Luddite movement. Unlike the bioethicist, however, who has authority granted to them by the state to enforce ethical oversight, Christianity has no such trump card against the transhumanists. While a doctor cannot ignore a bioethicist mandate without being punished by the law, transhumanism is under absolutely no imperative to listen to Christianity. If we criticize, we are villainized; if we approve, we become cheerleaders on the sideline… and all our beliefs are false and violent anyways. Or so the story goes.

The Current Predicament

There are already many various Christian stances toward transhumanism. In this volume we will read some authors who are critical of transhumanism, and others who are in support of the movement. They may appear to be complete opposites, but look closely because they share one crucial element in common: both are forced to play the bioethicist.[23] The Christians in this engagement thus far have to act as external ethical overseers, and they have thereby been preemptively stripped of all power. If they criticize transhumanism, all concerns can be easily ignored by shouting "progress!" loudly enough. If they give their approval, they will no longer serve a purpose in the conversation: they become a ticked checkbox in the process of ethical bureaucracy.

It is thus my conclusion that Christianity has lost a propaganda war—no matter what we conclude in the dialogue with transhumanism, we currently do not have the power to create any substantial change. And if Christianity is nothing more than a set of outdated beliefs about ethics, then why should we have any power to change the course of technological

[23] Of course it is possible to criticize or support transhumanism on other grounds and those can also be found in this book.

development in a conversation about science? But if Christianity is truly what it claimed to be in the Classical-Medieval days—a life of justice, love, and scientific reason in pursuit of humanity's union with the Trinity—then the entire transhumanist debate is fundamentally rigged against us and we need to ask ourselves why all the Christians are on the scientific sideline. How did "scientific progress" somehow become the mantra of atheism when it once belonged to Christianity?

In these last sections I will discuss two relatively recent trends in the natural sciences from which Christianity never recovered. Both occurred in Britain in the mid-1800s, and Christian denominations would be wise to deal with the divorce these historical problems caused as the first obstacle to overcome in the pursuit of reuniting Christianity and the natural sciences. Addressing these problems would open up new possibilities for Christians and help Christianity escape its rigged debate with transhumanism.

Divorcing Atheistic Science

As I discussed above, a palpable chasm had developed between Christianity and the natural sciences by the end of the 1500s.Yet this remained mostly theoretical for centuries. We might say that Christianity and the natural sciences were sleeping in separate rooms for these centuries. They were fighting, but the divorce papers had not been signed. From the 1500s up until the mid-1800s, the natural sciences were still flooded with openly devout Christians. Notice for example that three of the four pillars of modern science—Galileo, Boyle, and Newton—were all various kinds of Christian. (Laplace was an atheist, and considered eccentric because of it.) Even though many historians credit these minds with laying Enlightenment foundations for modern science that would ultimately conflict with Christianity, it took centuries for many people to realize consciously that the philosophical outlook of modern natural science was antagonistic toward the Christian philosophical outlook, even despite the perpetual march toward disproving the god-of-the-gaps. The fact that William Paley—now infamous for his "watchmaker god"—was required reading for all science students at Cambridge up until the 1850s is a surprising confirmation of this tension.[24]

[24] Harrison, *The Territories of Science and Religion*, 149.

The true "divorce" moment happened in Britain from the 1840s through the 1870s. Two pressures converged to purge the natural sciences of all Christian influence. First, the practitioners of the natural sciences needed to professionalize their discipline in order to guarantee quality control. This meant the removal of Christian clergy from the ranks of prestigious scientific societies.[25] This trend was confirming evidence to the likes of Max Weber and Francis Galton that "science today is irreligious"[26] because "the pursuit of science is uncongenial to the priestly character."[27]

Charles Darwin's *On the Origin of Species* was published in 1859, when this purge was well underway. Darwin's evolutionary theory was hailed as the final proof that no "god-of-the-gaps" existed, since materialistic selection pressures in the natural world could explain all biological change. For centuries post-Reformation culture had been searching for a god who visibly tinkered with the world in extraordinary ways.[28] For many, Darwin's theory meant no such god was necessary, and this left atheism as the only alternative. Darwinian evolution brought the theoretical chasm between Christianity and the natural sciences to the forefront of popular thought, creating a new culture war and marking the moment the "divorce" became final in the West's popular imagination.

Darwinian evolution and Christian theology did not necessarily have to conflict, and certain Christians in the wake of Darwin were able to harmonize the scientific data with Christian theology without distorting either. Generally speaking, however, there were two major reactions toward evolution in the Christian world, which would grow into an ideological rift. Evolution was widely believed to require an atheistic worldview, and the theory was hailed as a major success of the (newly professional) natural sciences. Thus popular imagination began to conflate evolution with atheism and science, which remains a common assumption in Britain and America to this day.[29] Many Christians have failed to question this conflation, but have responded in various ways toward it. One Christian faction became reactionary, essentially agreeing with Weber and Galton that science and religion are enemies, but choosing Christianity over the sciences. Some of these reactionaries evolved into today's fundamentalist

[25] Ibid., 159–164.
[26] Ibid., 159.
[27] Ibid., 250.
[28] Taylor, *Sources of the Self*, 272.
[29] See ibid., Chapter 22.

Protestant denominations. Other Protestants felt the march of "science" was impossible to inhibit or discredit; the only option was capitulation, even if it meant revising the core doctrines of Christianity itself. The resulting Christianity attempted to make itself relevant, scientific and modern, and evolved into today's liberal Protestant denominations.[30] The Fundamentalist tendency is to dismiss the atheistic scientist, the Liberal tendency is to accept the progressive notions of the atheistic scientist without reservation, but all too often both sides fail to ask why the scientist necessarily has to be an atheist in the first place. Such jaded fundamentalist attitudes toward scientific developments have convinced the transhumanists that Christianity is a Luddite religion. Such accommodating liberal attitudes toward scientific developments turn Christianity into the bioethicist religion that has nothing to offer transhumanism. To be blunt, Christianity dug its own grave in the transhumanist debate when we failed to arrive at a united, nuanced and reasonable stance toward the theory of evolution almost two centuries ago.

It is still critically important that Christians engage this divorce, because it has affected the educational structure of our universities and corroded the teaching ability of churches. Both of these effects serve to perpetuate the belief that Christianity and the natural sciences are divorced, thereby rigging the transhumanist debate before the sides can even meet.

The Idea of the Christian University

Universities guide the intellectual development of their culture, so we ought to pay special attention to their curriculum, structure, and method, since these are major influences on the intelligence and imagination of a society. It should therefore concern us that theology and the natural sciences are divorced in the typical university. Theology departments rarely interact with the scientific departments, even at Christian universities. Over time, the professionalization of the natural sciences inspired nearly every other academic discipline to professionalize. Constant professionalization over the decades has evolved into a cultural insistence upon specialization and sub-specialization, resulting in theologians who specialize in their own research, and natural scientists who specialize in theirs. Both learn to act as if the other discipline does not matter for their own work. Simply by growing accustomed to the standard structure of academia,

[30] See ibid., Chapters 14; 16.

theologians are conditioned to act as if God has nothing to reveal to them in the scientific exploration of the natural world and natural scientists are similarly conditioned to conclude that their pursuit is secular, or perhaps even militantly atheistic. Even in many Christian schools the only theological training that science students receive is from core introduction courses. They are not provided with opportunities to grow an articulate Christianity and weave it together with their work in the natural sciences. A similar dynamic exists with theology students and science classes.[31]

This departmental divorce has the consequence of making the theology students scientifically illiterate and the science students theologically illiterate. These students will go on to become the professors and experts and continue to drift apart. Such a scenario can happen at both Christian and secular schools. In the normal flow of modern academic life, theology and the natural sciences never have a chance to get to know each other. Max Weber's and Francis Galton's visions of anti-religious natural sciences are fulfilled daily in the structures of our universities.

Thus it seems that part of the reason for the continued divorce between Christianity and the natural sciences is because they are divorced in the universities that are the intellectual epicenters of our culture. If our educational method sends the message to students that there is a divorce between Christianity and the natural sciences, it should not be surprising when that generation grows up and imagines the two to be at war. Thus, Christian universities can begin changing Christianity's "bioethicist predicament" by rethinking the dominant educational model.

Christian universities might be wise to revive classical liberal arts education for this reason. The liberal arts curriculum stresses that each discipline, including the natural sciences, ought to lead to moral improvement and contemplation of God and creation. Since theology is the science focused most on contemplation of God and creation, the liberal arts curriculum has already reserved an important space for theology, and an established tie between Christianity and the natural sciences already exists there. The classic liberal arts curriculum already implicitly begins to weave Christianity and the natural sciences back together. Even if a return to the liberal arts educational model is not the proper fit for a certain school, it is a time-tested curriculum with many virtues which will provide educators with plenty of examples and ideas to consider and augment.

[31] See Gregory, *The Unintended Reformation*, Chapter 6.

Preaching the Natural Sciences

The divorce has also caused Christians leaders to neglect important subjects in their preaching. Christian preachers rarely encourage the natural sciences, and our Christian ancestors would consider this a serious missed opportunity. By failing to define work in the natural sciences as a Christian vocation, preachers send the same message to their congregation that universities send to their students: Christianity and the natural sciences are divorced.

But if the Classical-Medieval Christians were correct about the marriage between *religio* and *scientia,* then God provides grace through the rigorous study of the natural world, and today's Christians are missing out on this grace insofar as they avoid the scientific disciplines. Thus from the Classical-Medieval perspective, the failure of Christian preachers to encourage Christian engagement in the nature sciences is a moral failure that will inhibit the spiritual growth of the congregation. Christianity without the natural sciences is Christianity torn to shreds, at least according to Christians of past eras.

Perhaps nobody has stressed this point lately as much as the Rev. Dr. Christopher Benek. Dr. Benek recalls that the first time he preached to his congregation about seeing technological and scientific advancement as a form of Christian charity, a chemical engineer in his parish approached him after the service. The chemical engineer was grateful to the point of tears because Dr. Benek was the first Christian to proclaim that his engineering work was important to God. The chemical engineer had been doing this same work for decades, but his fellow Christians had always either villainized it or considered it "spiritually unimportant."

Dr. Benek's story illustrates the problem well. Modern universities and scientists are not the only ones to perpetuate the myth that Christianity and natural science are divorced—Christians perpetuate this myth too, and make it part of the culture of our churches. Again, this only provides transhumanists with more reasons to dismiss Christianity.

Christian denominations ought to instead make it their goal to re-baptize the natural sciences, by growing articulate Christians with a deep knowledge of both theology and natural science who will become the leading researchers and scholars in each discipline. The most effective way to undo the divorce is to make it so that the best scientists are the best Christians. Theologically literate Christians need to be the ones doing the avant-garde scientific and technological research, and they need to be the first to interpret the data in the theoretical sciences. Intelligent Christian theology ought to become so enmeshed with scientific discovery in practice that when the greater culture engages the science it is thereby forced to engage the best of Christian theology at the same time. If this were to happen the debate with transhumanism would look radically different. In fact, atheistic strands of transhumanism might die out, since scientific progress would become the reclaimed mantra of Christianity.

Here are a few suggestions for helping to create such reform: First, Christian leaders need to stress that being a scientist is a vocation, that work in the natural sciences can benefit one's soul and provide charity for those in need. As part of this, Christian leaders need role models in the sciences that the congregation—especially the youth—can model themselves after. There are plenty of Christian scientists to choose: Copernicus, Albert the Great, and Roger Bacon (not to be confused with Francis Bacon), are a few out of hundreds.

Second, the resurrection of feast days would be very helpful for turning these mostly-forgotten Christian scientists back into role models. Christians who were especially good role models were commemorated by parishes after their death with their own feast day each year. These feast days would be distinctive celebrations of their lives and usually included practices to teach the congregation about those lives. By making such celebrations unique and recurring events in the official church calendar, the parish helped the congregation to build habits that would allow them to resemble those role models. Christian denominations need something like feast days to inspire their congregations to embrace the sciences.

Third and finally, many Christians need to change their attitude toward the natural world and technology. This is mainly a worry I have about Protestantism due to some common theological themes in the Protestant world (although the issue affects other groups as well).[32] Put bluntly, Protestants have not come to a shared understanding of God's relationship to Creation. Due largely to the Fundamentalist-Liberal split, Protestants have not been provided with a shared understanding of God's relationship to human technology (in fact, much of the Protestant laity does not seem aware of the question yet). Protestant theology from the

[32] A 2014 Pew survey, "Religious Landscape Survey: Views about human evolution" may be instructive in this regard. The survey reported that 38% of evangelical protestants, 50% of historically black protestants, 66% of Catholics, 64% of mainline protestants, and 59% of Orthodox Christians believe that humans have evolved over time.

beginning had a worrisome tendency to think of God's relationship to the world as a divine watchmaker whose interactions with the world were always miraculous and against the regularities of nature.[33] Darwin was so scandalous to the Protestant world because his evolutionary theory seemed to prove this god never existed, or at least never actually interacted with the world.[34] This scandal created a divorce between God and the natural world that has not been reconciled in the minds of many Protestants. This becomes especially problematic if combined with other common doctrines from various denominations. Let us enter the imagination of a certain Christian church and see where their logic leads. For starters, this church sees the sacraments as purely symbolic reminders. This church also accepts the doctrine of total depravity, resulting in radical pessimism against the human mind to discover God's Truth. Combining these doctrines places us in a universe where it is difficult to reliably find God's presence. God cannot be present in the natural world, because modern science left Him nowhere to exist. God would not be found in the sacraments because these are symbolic reminders. And it seems we must conclude that God is not present in technology because the majority of humanity is depraved. Is it possible for a holy thing to emerge by the work of totally depraved human hands?

The Eastern Orthodox attitude toward the natural world and technology—which is revealed in their icons—provides Christians with the imagination to approach the issue from a new angle. Icons are handcrafted paintings of Christ that brings the believer into greater intimacy with Him. These icons are products of nature and technology. They are made with wood, pigment and precious metals. But these materials have to be harvested and scientifically prepared. Human hands must be trained for years in the art of iconography before they can produce an icon worthy of devotional use.

Yet the Second Ecumenical Council of Nicaea in 787AD holds that these pieces of wood painted by human hands are so filled with Christ that a kiss of the icon is a kiss received immediately by Christ Himself. "The honor which is paid to the image (icon) passes on to that which the image represents (Christ), and he who reveres the image reveres in it the subject represented." Every time an Orthodox believer kisses an icon—which is a central part of their worship—they affirm that God is present in the natural

[33] Taylor, *Sources of the Self*, 403–404.
[34] Ibid.

world and human technology. The Orthodox universe encourages believers to explore the world through science and technology; believers are assured through their worship practices that God will reveal Himself through these endeavors.

Or let us take the Catholic Eucharist for a Western example. The Eucharist is a product of the natural world and human technology. The grain grew from the Earth after being cultivated and had to be harvested then carefully prepared in order to become bread appropriate for Communion. Once it arrives at the parish it must be kept in vessels of gold and silver, designed specifically to guard its purity. Once the priest gives the blessing, this bread is transformed into the physical Body of Christ, despite originally being a humble product of nature and technology. The Catholic universe is one where God's fullness can dwell in the natural world and technology. Simply by receiving Eucharist, the Catholic is habituated to see a universe full of God's presence, waiting to be revealed through scientific discovery and technological progress.

In the Catholic-Orthodox universe the sacraments—revelations of God's literal presence in the physical universe—act as promises that God will reveal Himself in scientific study and technological development. One can appreciate how such an approach would deepen the Christian imagination. It would thereby allow the faithful to see science and technology anew, as channels of grace worthy of Christian engagement. Rather than lightly dismissing Orthodox icons and Catholic Eucharist as idolatrous, these practices can be reconceived as treasures in the Christian universe. Part of their value lies in their potential to help Christianity escape the "bioethicist dilemma" by re-sanctifying Christian visions of science and technology. The icon and the Eucharist challenge some deeply held doctrines across denominations, but they introduce a form of Christianity that can rise to meet the challenges raised by transhumanism.

Bibliography

Barbour, Ian G. 1997. *Religion & Science: Historical & Contemporary Issues.* San Francisco: HarperCollins.

Benedict XVI, Pope. 2006. Faith, Reason, and the University: Regensburg Address. *Libreria Editrice Vaticana,* September 12. https://w2.vatican.va/content/benedict-xvi/en/speeches/2006/september/documents/hf_ben-xvi_spe_20060912_university-regensburg.html

Burtt, E.A. 1924. *The Metaphysical Foundations of Modern Physical Science*. Amherst: Humanity Books.

Cavanaugh, William T. 2009. *The Myth of Religious Violence: Secular Ideology and the Roots of Modern Conflict*. New York: Oxford University Press.

Ferngren, G.B. 2014. *Medicine and Religion: A Historical Introduction*. Baltimore: Johns Hopkins University Press.

Gregory, Brad S. 2015. *The Unintended Reformation: How a Religious Revolution Secularized Society*. Cambridge, MA: Harvard University Press.

Harrison, Peter. 2015. *The Territories of Science and Religion*. Chicago: University of Chicago Press.

Hart, David Bentley. 2009. *Atheist Delusions: The Christian Revolution and Its Fashionable Enemies*. New Haven: Yale University Press.

Hawking, Stephen, and Leonard Mlodinow. 2010. *The Grand Design*. New York: Bantam Books.

Pinker, Steven. 2015. The Moral Imperative for Bioethics. *The Boston Globe*, August 1. https://www.bostonglobe.com/opinion/2015/07/31/the-moral-imperative-for-bioethics/JmEkoyzlTAu9oQV76JrK9N/story.html

Religious Landscape Survey: Views About Human Evolution. *Pew Research Center*, 2014. http://www.pewforum.org/religious-landscape-study/views-about-human-evolution/

Stark, Rodney. 2015. *A Star in the East: The Rise of Christianity in China*. West Conshohocken: Templeton Press.

Taylor, Charles. 1992. *Sources of the Self: The Making of the Modern Identity*. Cambridge: Cambridge University Press.

PART II

Confronting Transhumanism: From Hype to Hope

CHAPTER 6

The *Imago Dei* and the *Imago Mundi*

Michael Dickson

The Scope of This Discussion

The self-described "transhumanist" movement is hardly characterized by caution, or doubt:

> Bio-fatalism will increasingly be replaced by techno-can-do-ism—the belief in the power of the new technology to free us from the limitations of our bodies and minds… In the twenty-first century, the belief in the Fall of Man will be replaced by the belief in his inevitable transcendence – through Superbiology.[1]

And yet, for reasons that many have pointed out, and have nothing, directly, to do with religious conviction, there are plenty of reasons to worry about the prudence of pursuing the transhumanist dream. Consideration of our limited ability to understand the consequences of technology, the fallibility of our implementation of technology, the

[1] Young, *Designer Evolution*, 20.

M. Dickson (✉)
Department of Philosophy, University of South Carolina, Columbia, SC, USA
e-mail: dickson@sc.edu

S. Donaldson, R. Cole-Turner (eds.), *Christian Perspectives on Transhumanism and the Church*, Palgrave Studies in the Future of Humanity and its Successors,
https://doi.org/10.1007/978-3-319-90323-1_6

potential for social injustice, and more, should give *anybody* reason to be concerned about the transhumanist vision. (As Peters[2] points out, these concerns might be made apparent through reflection on one's religious convictions.)

I emphasize at the outset that the point is not a form of Luddism. The considerations just mentioned are *always* important when one is entertaining a new technology, regardless of one's general level of enthusiasm about technological change. They become serious and potentially overriding considerations when the technology portends the most serious of consequences, should it go wrong.

For the Christian, there is a very specific, and very serious, way in which transhumanist proposals could go wrong. My aim is to elucidate that danger and to explain why it makes transhumanist technologies categorically different from other technologies (such as, say, nuclear technology) about which one might have prudential concerns, or feel a need for caution. Moreover, my remarks are not directed at the philosophical trappings of the transhumanist program, which often include a blend of social Darwinism, ethical consequentialism, and some version of (typically left-wing) libertarianism. My question will be further narrowed, in three respects.

First, significant extension of the normal human lifespan has been a focal point of the transhumanist movement, and so I will focus on it. Of course, transhumanists have entertained enhancements of physical, emotional, and cognitive capacities that are not directly related to extended lifespan (though might make extended lifespan more tolerable). These proposals range from adaptation of currently available technology for perhaps minor enhancements (e.g., using therapeutic drugs for the purpose of enhanced memory or concentration), to the development of radical new technologies (e.g., interfacing a brain with a computer), to the seemingly ridiculous.

Second, I will set aside many aspects of a Christian worldview that might lead one to adopt this or that stance toward these proposed technologies. For example, Keenan[3] suggests that the Christian understanding of salvation and eternal life with God is incompatible with the transhumanist version of

[2] Peters, "Progress and Provolution Will Transhumanism Leave Sin Behind?" in Cole-Turner, *Transhumanism and Transcendence*, 63–86.

[3] Keenan, James F., "Roman Catholic Christianity—Embodiment and Relationality: Roman Catholic Concerns about Transhumanist Proposals," in Mercer and Maher, *Transhumanism and the Body*, 155–172.

technologically-assisted "immortality." My focus is on the Christian teaching that humans were created in the "image and likeness of God."

Finally, I consider proposed technologies without reference to the explicit philosophical agendas that often accompany or even motivate them. However, for the record, I do not presume that the two are in fact independent. Indeed, I view the association between the proposed technologies and the philosophical commitments that frequently accompany those proposals as largely non-accidental. That agenda typically includes atheism and an understanding of the nature and importance of human flourishing, freedom, and autonomy that is at odds with almost any Christian understanding of those aspects of the human condition.

Setting aside any critique of that agenda, a serious question for Christians is this: To what extent can, or should, Christians endorse pursuing the sorts of technologies that are proposed by transhumanists, not in the context of *their* worldview, but in the context of a *Christian* worldview? The remainder of this chapter considers that question, with special reference to radical extension of human lifespan and the Christian teaching that we are created in the image of God. In the next section, I consider and reject an argument that is sometimes made that the intrinsic value of life provides the Christian a reason to promote (or at least accept) radical extension of lifespan. (I have no very specific definition of "radical" in mind. As a ballpark figure, let us suppose that any extension of more than, say, 50 years, counts as "radical." That ballpark figure will do, inasmuch as transhumanists typically have a much more ambitious goal in mind.) The subsequent section provides a brief overview of various ways of thinking about the Christian teaching that humans are created in the image of God. The remainder of the chapter considers the effect that this doctrine might have on the Christian assessment of transhumanist goals and proposals, especially as regards radical extension of human lifespan.

A Supposedly Christian Argument for Extension of Lifespan

Some[4] have argued that the Christian emphasis on the intrinsic value of life comports well with the transhumanist goal of longer or even unlimited lifespan. Such arguments are often accompanied by the following bit

[4] E.g., Max More, "Why Catholics Should Support the Transhumanist Goal of Extended Life," in Vaccaro, *L'ultimo esorcismo*.

of sophistry: the average lifespan of pre-historic humans was somewhere around 30 years; everybody sees our current average lifespan of around 70 years as an improvement; therefore, additional extensions would be an additional improvement. For example, Bostrom[5] writes: "Average human lifespan hovered between 20 and 30 years for most of our species' history. Most people today are thus living highly unnaturally long lives," the intended implication being that living an *even more* "unnaturally" long life is good.

Apart from the fact that this argument relies on the false principle that "if one increase of *X* is good, then further increase of *X* is better," it also relies on a pernicious misrepresentation of the facts. Yes, the average life expectancy even a century ago was shorter than it is now, by any reasonable reading of the evidence. However, this statistical fact reflects not an extension of the natural upper bound on human lifespan, but an increased capacity to prevent and treat formerly lethal misfortunes. Until recently, the steady increase in life expectancy was almost entirely due to the reduction of infant mortality although, very recently, advances have also been based on our ability to keep middle-aged and older people alive longer by successfully treating diseases that previously would have killed them. There is no clear evidence that the age at which people tend to die due to general decay of bodily function has changed in the course of human history. (Plenty of ancients lived to 70 and well beyond, for example.)

Thus the argument is flawed but, worse, it is dangerous. It is dangerous because it encourages the thought that the Christian valuing of human lives is connected with the length of those lives. However, from a traditional Christian perspective, the *value* of a human life does not depend on its length.

To illustrate, the value of the lives of an aborted fetus, a young lady who died in an accident, and an elderly man who died of natural causes, is the same in each case. One may, in some of these cases, woefully regret the circumstances of death, including its timing and what might had been had it not occurred. One may thus judge that the circumstances were in some cases unfortunate, or even evil, but those judgments have no bearing on the value of the life that was lost, which does not depend on the length of the life.

[5] Bostrom, "Introduction—The Transhumanist FAQ: A General Introduction," in Mercer and Maher, *Transhumanism and the Body*, 11–17.

Of course, there may be important qualitative differences among the three lives that I mentioned above, and the relative lengths of the lives might even be relevant to those differences, but they do not differ in intrinsic value. The flip side, of course, is that we do not make a life "more intrinsically valuable"—indeed, we do not necessarily even make it "better" in any sense—by making it longer. There is, of course, a hornet's nest of issues (abortion, capital punishment, birth control, etc.) in the vicinity of these ideas. Here I make only one simple point: The Christian insistence on the intrinsic value of life does not entail that longer life is more valuable (nor does the insistence that longer life is *not* necessarily more valuable life qualify one as a "deathist," a silly and undefined term used by some transhumanists to vilify those who take a stance against their goals). Indeed, the specifically Christian manner of valuing life entails that all lives are equally valuable.

Now, one might agree that by making a life longer, we do not thereby increase its intrinsic value, but nevertheless claim that (all other things being equal) the life of the elderly man was more fulfilled, while the life of the young lady was in a significant way "cut short," and the fetus was not even given a chance to develop its full human capacities. Hence there is an important sense in which these lives differ in their quality, if not their intrinsic value. However, it does not follow from this observation that extending human lifespan further improves that quality. The point is not that from age 100 to 150 one might live a miserable existence of frailty and disease—transhumanists certainly do not endorse *that* goal—but simply that longer life is not always better. It might be better for the fetus and the young lady. It might not be better for those who live what we would now consider a "full" life. Indeed, there are very *few* goods that are always better in greater quantities and, in any case, there is nothing (that I can see) in the traditional Christian insistence on the *value of life* that implies that *extending human lifespan* is a good, any more than extending human access to other goods (such as food and drink) is necessarily a good.

There is plenty more to examine. What, for example, of the resurrection? Does it not imply infinitely extended life, and is that life not good? It is, but the good of that life, in which we are "like the angels in heaven,"[6] is not a good that can be granted by any technology, or so I shall assume here. Here I am focused on the kinds of life-extension that could be achieved by human technology, and the claim is only that more of *that*

[6] Matthew 22:30.

sort of life is not *necessarily* better. Below I will raise some more specific worries about the notion that longer life is always better, focusing on those that arise from the Christian teaching that we are created in the image of God, to which I now turn.

The Imago Dei: How and What

The teaching that we are made "in the image and likeness of God" (imago Dei) has been a source of analysis, reflection, puzzlement, and debate since the days of the early Church Fathers. Of necessity, I set aside a number of important distinctions that have been made in the pursuit of understanding it, in order to simplify the discussion—I hope in ways that are illuminating rather than simplistic or misleading. I will attend to two distinctions.

The first concerns *how* we are images of God. It is common to divide answers into three types.[7] According to the "structural" (sometimes, "substantial") view, the image of God resides in some aspect or aspects of human nature. According to the "functional" view, the image of God resides in a capacity, or exercise of a capacity, somehow reflecting God's capacities. According to the "relational" view, the image resides in personal relationships, either between humans or between humans and God.

A structural version of the teaching, which has arguably been the dominant view at least until recent times, focuses on human nature itself, seeing it as somehow like, or an image of, God. For example, one common view is that we are like God inasmuch as we are rational—indeed, Cairns remarks, perhaps exaggerating, that "in all Christian writing up to Aquinas we find the image of God conceived as man's power of reason."[8]

A functional view of the teaching focuses on human capacities, seeing some of the acts that those capacities allow us to perform as similar in some way to the acts that God performs. In recent decades, the most commonly defended functional view—most especially among Old Testament scholars—is that we are like God inasmuch as we have the capacity to exercise dominion over other parts of creation. (See, e.g., Barr,[9] who does not himself subscribe to the view, but describes it as "the most influential

[7] This division follows Erickson, *Christian Theology*, 520–527. Cf. Middleton, *The Liberating Image*.

[8] Cairns, *The Image of God in Man*, 110.

[9] Barr, *Biblical Faith and Natural Theology*, 158.

opinion today" on the matter.) This view is bolstered (according to its proponents) both by consideration of the historical context of the text (which I cannot consider here) and the proximity, in *Genesis* 1:26, of God's deliberation to create humans to His deliberation to allow their "rule" over His creation.

A relational view of the teaching focuses on human relations, and, in its most prominent form, the relation between man and woman. Barth, who is perhaps the most famous exponent of this view, is motivated by the use of the plural in *Genesis* 1:26 ("let *us* make man in *our* image, in *our* likeness"), and by the explicit specification of both male and female in *Genesis* 1:27 and *Genesis* 5:1–2. He suggests that the image thus consists in the "juxtaposition and conjunction of man and man which is that of male and female,"[10] which is a reflection of "the relationship and differentiation between the I and the Thou in God himself" manifested in "the relationship of man and woman in which man is a Thou to his fellow and therefore himself an I in responsibility to this claim."[11] (A clearer, if less specific, statement of the view is: "God created him in His own image in the fact that He did not create him alone but in this connexion and fellowship. … God is in relationship, and so too is the man created by Him. This is his divine likeness."[12])

For the sake of full disclosure I agree with Aquinas and the tradition that the structural view is the fundamental one. Whatever is correct about the others follows from it rather than being themselves constitutive of the imago Dei. Indeed, the non-structuralist views are often (though not, perhaps, necessarily) motivated by philosophical commitments that are in tension with the Christian tradition. For example, the relational view is often motivated by a commitment to some form of existentialism, according to which there is no such thing as "human nature." In what is perhaps one of the more stark expressions of this view, Soltoveitchik writes:

> 'To be' means to be the only one, singular and different, and consequently lonely. For what causes man to be lonely and feel insecure if not the awareness of uniqueness and exclusiveness? The 'I' is lonely, experiencing ontological incompleteness and causalness, because there is no one who exists like the 'I' and because the *modus existentiae* of the 'I' cannot be repeated, imitated, or experienced by others.[13]

[10] Barth, *Church Dogmatics*, §3.1.195.
[11] Ibid., §3.1.198.
[12] Ibid., §3.2.324.
[13] Soltoveitchik, *The Lonely Man of Faith*, 39–40.

Barth, in a characteristically more careful and subtle manner, nonetheless makes similar assertions, saying that there is nothing *about* human beings—i.e., nothing about so-called "human nature"—that makes us images of God.

Proponents of the functional view are similarly inclined to downplay the role, or even to deny the objective reality, of "human nature." Clines,[14] for example, proposes a version of the "dominion" variant of the functional view, describing it thus:

> The image is to be understood not so much ontologically as existentially: it comes to expression not in the nature of man so much as in his activity and function. This function is to represent God's lordship to the lower orders of creation.[15]

Clines takes humans to be "representative rather than representation" of God; i.e., we are "images" not because of a relation of similarity that we bear to God, but because we were created by God as his representatives, to exercise dominion. Dominion is thus "a constitutive part of the image itself."[16] (Clines does not seem to want to follow existentialists in denying such a thing as "human nature," however.)

Quite apart from the fact that a denial or downplaying of the role of human nature in defining our relationship to God clashes with the history of philosophical and theological thought throughout most of the Christian tradition, it faces two other problems, one textual, the other explanatory.

The textual problem, explored in some detail by Gardoski,[17] arises from the fact that while functional or relational features of God's creation are indeed described in close proximity to the description of His creation of human beings, the natural reading of the Hebrew appears to be that they are described either as consequences of that creation, or as directives from God to the created beings.

The explanatory problem arises from the question how we are to explain the human capacity for dominion (or for the exercise of *any* function) if not in terms of some aspect of human nature. The most natural explanation of our capacity for dominion is in terms of our being a particular kind of creature, whose nature gives us the capacities. Similarly, the

[14] Clines, "The Image of God in Man."
[15] Ibid., 101.
[16] Ibid., 96.
[17] Gardoski, "The Imago Dei Revisited."

natural way to understand relationships is as arising from the characteristics ("natures") of the persons that enter into those relationships. Both sorts of explanation suggest that we are images of God because God created us to be beings with a particular nature, and it is *in virtue of* that nature that we have certain capacities and that we enter into certain relationships. Nonetheless, I will keep all three views in play, for the time being.

The second distinction to which I will attend concerns the question, "What *exactly* is made in the image and likeness of God?" The distinction plays out differently for each of the different views, but the general contours of the possible answers are the same in each. In each case one speaks of "body," "soul," and "complex of body and soul." The last case may embrace an anti-dualistic view in which body and soul are somehow unavoidably united, perhaps distinguishable only conceptually but not in reality.

On a structural view, the potential answers to this question have traditionally been that the image of God resides in the thing itself, either soul, body, or the complex of body and soul. Perplexities attend each answer. If the body alone, then are we to say that God too has a physical form (contrary to scripture)? If the soul alone, what do we say to Old Testament scholars,[18] who often remark that the word used (*selem*) naturally refers to a *physical* image? And why, when God said "Let us make man in our image, after our likeness," did that creation have both a body and a soul, and not just a soul? On the other hand, if we are images in virtue of the complex of body and soul, then is God Himself somehow a complex of "body" and "soul"? Does that conception of God contradict His simplicity (not to mention Biblical passages that seem to affirm that God is pure spirit)?

Similar issues arise for the functional and relational views. What sort of function is it that is an image of God? A function of the body alone? The soul alone? The complex of body and soul? Similarly, what sort of relation? Barth, for example, puts strong emphasis on the relation between man and woman, and puts some stress on the physical union of husband and wife, which is at least partly bodily in its nature.

None of these questions lacks sophisticated replies in the philosophical and theological literature. My initial point is simply that the question "What *exactly* is made in the image and likeness of God?" has many possible answers, each of which seems to raise more questions. I will suggest below that the most plausible form of Christian transhumanism should

[18] E.g., Middleton, *The Liberating Image*, 45ff.

adopt the view that the image of God resides entirely in the soul (though I myself think that it more likely concerns the complex of body and soul).

A further issue is the effect of the Fall on the image of God. (Here I intend to speak of "the Fall" broadly, to include the view that it was an historical event and also less literal views; the real point is to understand the sense in which we are "fallen," and the effect of that fact on the "image of God.") Regardless of one's view here, the New Testament seems to indicate that the image of God is *somehow* present in humans, in some manner that is relevant to action. For example, *James* 3:9 and especially *1 Corinthians* 11:7 speak of humans made in the image of God in a manner that indicates the relevance of this fact for human action. At the same time, some Biblical texts do emphasize the radical effects of the Fall. Most theologians conclude that either the image is damaged or obscured in some manner, but not completely lost, or that the image has multiple (normally, two) aspects, one of which is completely lost, the other of which remains. For our purposes, as important as these issues are, all that matters is that there is *something* in us of the image of God. As the New Testament passages indicate, that fact is enough to guide action in at least some respects, and I take this view as given for the remainder of the analysis.

The Imago Dei and Transhumanism

The previous section left us with various forms that an explication of the teaching of the imago Dei could take. I have indicated what I consider to be a general outline of the most likely form (that it is structural and concerns the complex of body and soul). However, for present purposes I set that view aside, and consider what form the teaching of the imago Dei might best take, if it is to be amenable to some version of the transhumanist agenda.

The "trans" in "transhumanism" refers to the idea that the intended long-term outcome is to "eventually manage to become posthuman,"[19] something *other than* human. The *explicit* aim of the technology is to overcome, or destroy, human nature, in favor of something "better." This aim puts transhumanist technology in a different category from other technologies, even those with potentially serious consequences, and it presses upon Christians the responsibility to understand whether, by adopting transhumanist technologies we thereby destroy, or further damage, the image of God. The previous section should help us in this task.

[19] Bostrom, "Transhumanist Values," 4.

Note that the specific technologies proposed by transhumanists are inevitably bodily in their implementation. No transhumanist is proposing "direct manipulation of the soul" (as might be said to be effected by God, for example, via the administration of holy sacraments, or through prayer). The safest form of "Christian transhumanism," therefore, is one according to which the "what" of the imago Dei concerns the soul alone (whether in its nature, its functions, or its relations). In that case, one could argue that making the body *other* than what it is, that is, changing its very *nature* and thereby making it *other* than what God created, would not damage the image of God.

A plausible version of Christian transhumanism, then, will be one that answers the "what?" question with the soul alone, or some function or relation of the soul alone (if there is such a thing). Indeed, on a functional or relational view, transhuman alteration of the *body* might seem to be potentially a good thing. For example, if exercising dominion over God's creation somehow reflects God's perfect dominion, then exercising *better* dominion, perhaps by virtue of some alteration to our bodies, could *better* reflect God's perfect dominion.

Similarly, if an alteration to the body were to enable the relation between husband and wife to be *more* like the relations among the persons of the Trinity—for example, *more* harmonious in some fashion—then again, perhaps, that alteration would *not* damage, and might even *enhance*, the image of God in us.

If all of the above is correct, then a transhumanism that does not portend damage to the imago Dei could take one of three (not mutually exclusive) forms:

1. Alteration of the nature of the body that has no effect on or "improves" the soul (structural transhumanism)
2. Alteration of the nature of the body that has no effect on or "improves" the soul's functions (functional transhumanism)
3. Alteration of the nature of the body that has no effect on or "improves" the relations into which the soul enters (relational transhumanism)

I proceed by considering the example of radical lifespan extension in each case, raising doubts about the viability of any of these sorts of alteration. For present purposes, my only aim is to raise serious doubts.

Structural Transhumanism

Would radical lifespan extension have no effect on the soul? The claim seems inherently implausible. What, for example, would be the effect on one's tenacity in pursuing projects and plans? If life became very easily maintained, what would be the effect on our valuing of life? After having spent, say, 50, or 80 years, exploring what life has to offer, new ideas, and so on, would we become complacent?

Here is the report of one gerontologist: "I live among affluent, elderly people between the ages of 70 and 95. They are in good health, they have money, and they can take nice cruises or just putter about. They go to Scottsdale and play golf. But they don't seem to have any new energy, and they sure don't have any new serious initiatives."[20] The point is not that these individuals are decrepit or miserable, but that their outlook on life has changed significantly. Whether that change is "good" or "bad" is not the issue here (and of course we should not presume that everybody is or would be as described). The point, rather, is that it seems likely that if we lived half or more of our lives in this manner, we would indeed become very different sorts of people. In a similar vein, if teenagers, or even 20-somethings, or 30-somethings, lived in a world where everybody lived to 170, how eager would they be to start families, to pursue new projects, and to improve themselves? After all, they have all the time in the world to do those things. Can we rest assured that *none* of these changes, or others, would occur in the face of radically extended life?

More generally, it is difficult to deny that our environment, and especially our social and family environment, affects how we view and engage with the world. Now, those differences are not, perhaps, differences in the nature of souls (though perhaps they are), but what do we say about an entire society, or the entire human race, raised, generation after generation, in an environment of peers who are radically different from how they have been at any point in our species' history? Would our souls somehow remain unchanged in the face of these differences in society and family? It is not clear.

Functional Transhumanism

Even if, somehow, the nature of the human soul remained constant in the face of radical extension of human lifespan, it is likely that its ability to

[20] Stock and Callahan, "Point-Counterpoint: Would Doubling the Human Life Span be a Net Positive or Negative for Us either as Individuals or as a Society?"

carry out its natural functions would change and, either immediately or over time, it is entirely possible that those natural functions would themselves change. In the case of some transhumanist proposals, this fact is obvious because the very *goal* of the technology is to enable "better" functioning (effortless recall of memories, effortless and speedy computation, absence of inhibiting emotions, and so on) or to produce entirely new functions (e.g., sensory). However, even in the case of extension of lifespan, which is not explicitly aimed at changing our capacities, it is far from clear that the effect would not be to alter both the capacities that we have and the facility with which we can exercise them. Would extremely long-lived people eventually simply grow weary of loving their families? Their neighbors? Would they grow weary of charity? Or would they perhaps become fixated by it to the detriment of the exercise of other capacities?

Predicting these outcomes is extremely difficult—and some of them could be *positive* for the Christian (my aim is not to deny this possibility). We are treading on unexplored territory. What is less difficult to predict, given that the development of the soul depends in many ways on the body, is that a radical change in the body, including radical life-extension (which, after all, will come only in the wake of *other* radical changes, such as greater control over the natural processes that currently govern the body), will have *some effect or other* on the soul's functioning.

Relational Transhumanism

Perhaps the most obvious changes that would occur in the face of radical lifespan extension would be in the relationships that we bear to one another. The most obvious cases involve the family, and especially the institution of marriage. How would an average lifespan of, say, 150 years bear on the relationship that married couples share? Would people wait until their 40s, or 50s, or 100s, to get married? How would that change affect the normal relationship between husband and wife, or parents, grandparents and children? I do not know the answers to these questions, but the key point is that there would in fact be a significant effect. After all, human relationships are, on the relational view of the imago Dei, the essence of being human.

The general claim of this section is just that radical extension of lifespan is *very* unlikely to leave the soul, its functions, and its relations, alone. Moreover, while we can speculate on what changes would occur, it is extremely difficult to do so with any significant degree of confidence.

A further brief point: I have focused on the bare fact of life-extension, but it must be emphasized as well that the means by which it is brought about are far from irrelevant. This observation raises two important, final, points to make, here. First, those means will necessarily involve a degree of control over our bodies, and a degree of intervention in bodily processes that is unprecedented in human history and that will almost surely alter the manner in which we think about our bodies and more generally our relationship to the physical world. Second, those means might themselves, by Christian standards, be immoral, in which case they are illicit no matter the outcome. They might involve an inherent disrespect for life, for example.

To summarize, given that transhumanists seek literally to change the underlying nature of the human being by changing the body, the most plausible way to reconcile transhumanist goals with the teaching that we are made in the image of God is to locate that image in the soul, whether in the soul itself, its functions, or relations. However, even in that case, it seems very likely that radical change of the body, especially radical extension of lifespan, would have some substantial impact on the soul, its functions, and relations. Moreover, the means by which those changes might be effected are potentially contrary to Christian morality.

An Objection and the Imago Mundi

To the non-Christian transhumanist, the argument of the previous section as summarized in its last paragraph is beside the point. Non-Christian transhumanists need not care about the teaching of the imago Dei, and so may acknowledge that the changes to the body that they propose will have consequences for whatever in their worldview passes for the "soul." Indeed, in many cases, such changes are the *goal*.

Often the transhumanists have little patience for the warning that those consequences may be negative. While not always ignoring that possibility, transhumanists complain about pessimists who see only "negative" consequences of transhumanist technologies. For example, after dismissing moral concerns about genetic enhancement of babies, and then rehearsing some real worries about the potential damage to the relationship between parent and child, Bostrom complains that "such dystopian scenarios are speculations" and suggests that "[w]e might speculate, instead, that germ-line enhancements will lead to more love and parental dedication," because, after all, "[s]ome mothers and fathers might find it easier to love a child who, thanks to enhancements, is bright, beautiful, healthy, and happy."[21]

Similarly, one might object that I have neglected the very possibility that some hold most dear, namely, that by altering the body in some substantial manner, they hope to *improve* the image of God by *improving* the soul, its functional capacities, or its relations. In this vein, transhumanists (and not merely Christian transhumanists) speak of improvements such as faster or more reliable cognition, greater emotional stability, freedom from emotional pain associated with disease, and so on. They also speak of improvements to the development of human relations, including, potentially, our relationship to God. (Max More, not himself a Christian, says that longer lifespan would offer "an extended opportunity to improve ourselves, do good works to redeem ourselves, to glorify God, and to more fully earn a place in Heaven."[22] Of course, More ignores the Biblical injunction to be prepared for Christ's coming at all hours, and not just after a conveniently long opportunity to get it right.)

In fact, the Fall provides just the space that might be needed to make this case. If the Fall damaged the imago Dei, then it can be repaired, and thereby the argument of the previous section is turned on its head: Yes, bodily alterations *can* affect the soul, and they can do so in a way that constitutes not further damage to the imago Dei, but an improvement or restoration of it.

But who knows what the true effects of these technologies will be, and who knows whether those effects truly constitute a (partial) restoration of the imago Dei, or instead a deformation of it? It is precisely *because* of the Fall that we should be wary of our ability to make that determination.

Both points work together. It would be easy to describe changes to the human condition in a manner that would constitute, if not a restoration of the imago Dei, at least a step in the right direction. Bostrom suggests that we consider describing the outcome of genetically enhanced babies as something like 'greater parental love.'[23] But why think it a *real* enhancement of love instead of (as, frankly, seems more likely) the appearance and trappings of love founded, however, on pride (in having produced such a

[21] Bostrom, "Human Genetic Enhancements: A Transhumanist Perspective."

[22] More, Max, "Why Catholics Should Support the Transhumanist Goal of Extended Life," in Vaccaro, *L'ultimo esorcismo*.

[23] Bostrom, "Human Genetic Enhancements: A Transhumanist Perspective."

wonder) or laziness (from not having to worry about the child), or some such sin? *True* parental love does not diminish in the face of a lack of beauty and health. Indeed, is it even *possible* for most humans to develop such true love of children, love that would withstand sorrow and pain, if there never were sorrow and pain? I do not presume to know the answer to this question, but I do suggest that nobody knows that the answer is "yes." We simply do not understand ourselves well enough to know, and it seems quite likely that one of the most powerful noetic effects of the Fall is that we never will, at least not while in our fallen state.

In short, how can we be confident that in the all too human pursuit of transhumanist improvement of the imago Dei, we will not wind up substituting our own, fallen, vision of goodness for God's, producing an image that is "of the world" (an "imago mundi")? Making ourselves into an image of what the world holds up as the good and the right, or what we might be tempted to take to be the good and right, is a serious risk in this endeavor, with the most serious of consequences.

Some transhumanists[24] have begun to claim that it might be immoral or criminal, to prevent, or even to avoid, research into radical lifespan extension or into other transhumanist goals. Bostrom, for example, after exploring a fable that portrays death as an unnatural event imposed on innocent victims, concludes that, "Searching for a cure for aging is not just a nice thing that we should perhaps one day get around to. It is an urgent, screaming moral imperative."[25] Assuming that aging *can* be prevented—and despite the rhetoric of transhumanists, this question remains open—there is an unqualified moral imperative to do so *only* if preventing death were always and under all circumstances the highest moral priority.

In contrast, no Christian could think that preventing death is always and under all circumstances the highest moral priority, for at least one very obvious reason—Christians are obligated to consider the entire picture. I have suggested that a crucial aspect of the entire picture is that the transhumanist goal of changing human nature *could* amount to damaging or destroying whatever in us, or about us, is an image of God. Moreover, we are not well placed to determine whether it will, and might even be predisposed to make poor judgements about this question.

[24] E.g., Zoltan Istvan, "When Does Hindering Life Extension Science Become a Crime?"
[25] Bostrom, "The Fable of the Dragon-Tyrant," 277.

Conclusion

I conclude that Christians must be extremely cautious when considering the advisability of pursuing the technologies proposed by transhumanists. This need for caution is not merely a species of the usual prudence that should attend the pursuit of new technologies, but is based on three considerations. First, transhumanist technologies avowedly aim to be not restorative but transformative of human nature itself, which makes these technologies a significant potential threat to the imago Dei. Second, we are particularly ill suited to predict the true consequences of implementing such ambitious technologies, even with the best of intentions, because doing so depends on having a very clear understanding not only of physical law, but also of our own nature and, most specifically, what it is about us that constitutes an imago Dei. Third, and related to the second, it might be very tempting, especially in our fallen state, to substitute what is expedient or strongly desired or pleasurable for the true imago Dei, thereby making ourselves into a kind of "imago mundi" instead.

None of these cautions should lead Christians to presume that the *specific* technologies proposed by transhumanists are, lock, stock, and barrel, unacceptable. They must be evaluated on a case-by-case basis and, in many cases, they might be found to serve the interests of Christians quite well, when pursued well—they might be found to reduce suffering, or to enhance life in some fashion or another. However, the form that the evaluation takes, for the Christian, must be radically different from the form that it takes for the average transhumanist. Indeed, it must be the *opposite*, in the following sense. The transhumanist seeks to change us into something that we are not, and presumes to know what that something should be, and how to achieve it.

For the Christian, insofar as human nature, in its current form, is what God intended at creation, altering it is contrary to God's will and therefore illicit. Insofar as human nature, in its current form, represents a departure from what God intended as a result of the Fall, we are not in a position to say with confidence which changes would count as "improvements" and which would not. Moreover, ultimately we *cannot* be the agents of the restoration of the imago Dei. This claim should not be construed as a form of quietism about alleviating human suffering or pursuing human goods. It is only the specific claim that repairing the damage done by the Fall is not within our power. As Peter, speaking of Christ, teaches us:

"There is no salvation through anyone else, nor is there any other name under heaven given to the human race by which we are to be saved."[26]

Bostrom tells a rousing tale of a king who slays the horrible "Dragon-tyrant" ("death") and saves his people from a great evil.[27] But Bostrom's king had no reason for doubt about the morality of his efforts, and was faced only with the possibility of failure, not the possibility of acting contrary to God's will.

Here is a different tale. In the middle of the Garden there are two trees. Of one, Adam has been told by God, "From that tree you shall not eat; when you eat from it you shall die." Alas, when Adam now tries to remember which tree it was, he cannot tell them apart. Nor can Eve. Egged on by a serpent, Adam and Eve can see that the fruit of the trees is good for food and pleasing to the eyes. And after all, only *one* of the trees is forbidden. The other should be fine. What should they do?

Bibliography

Barr, James. 1993. *Biblical Faith and Natural Theology*. Oxford: Clarendon.

Barth, Karl. 1958. *Church Dogmatics*. Edinburgh: T & T Clark.

Bostrom, Nick. 2003. Human Genetic Enhancements: A Transhumanist Perspective. *Journal of Value Inquiry* 37 (4): 493–506.

———. 2005a. The Fable of the Dragon-Tyrant. *Journal of Medical Ethics* 31 (5): 273–277.

———. 2005b. Transhumanist Values. *Journal of Philosophical Research* 30 (Supplement): 3–14.

Cairns, David. 1953. *The Image of God in Man*. New York: Philosophical Library.

Clines, D.J.A. 1968. The Image of God in Man. *Tyndale Bulletin* 19: 53–103.

Cole-Turner, Ronald, ed. 2011. *Transhumanism and Transcendence: Christian Hope in an Age of Technological Enhancement*. Washington, DC: Georgetown University Press.

Erickson, Millard J. 1999. *Christian Theology*. Grand Rapids: Baker.

Gardoski, Kenneth M. 2007. The Imago Dei Revisited. *The Journal of Ministry and Theology* 11 (2): 5–37.

Istvan, Zoltan. 2014. When Does Hindering Life Extension Science Become a Crime? *Psychology Today*, January 31. https://www.psychologytoday.com/blog/the-transhumanist-philosopher/201401/when-does-hindering-life-extension-science-become-crime

[26] Acts 4:12 (New American Bible, Revised Edition).

[27] Bostrom, "The Fable of the Dragon-Tyrant."

Mercer, Calvin, and Derek F. Maher, eds. 2014. *Transhumanism and the Body: The World Religions Speak*. New York: Palgrave Macmillan.
Middleton, J. Richard. 2005. *The Liberating Image: The Imago Dei in Genesis 1*. Grand Rapids: Brazos Press.
Soltoveitchik, Joseph B. 1992. *The Lonely Man of Faith*. New York: Doubleday.
Stock, Gregory, and Daniel Callahan. 2004. Point-Counterpoint: Would Doubling the Human Life Span Be a Net Positive or Negative for Us Either as Individuals or as a Society? *Journals of Gerontology: Series A* 59 (6): B554–B559.
Vaccaro, Andrea, ed. 2009. *L'ultimo esorcismo*. New York: EDB.
Young, Simon. 2006. *Designer Evolution: A Transhumanist Manifesto*. Amherst: Prometheus Books.

CHAPTER 7

Nietzsche's Power Ontology and Transhumanism: Or Why Christians Cannot Be Transhumanists

Jeffrey P. Bishop

Introduction

When living in a society that holds to a modern social imaginary—as Charles Taylor has called it—the utilitarian moral assessment of technology seems irresistible. By modern social imaginary I mean the shared values that set out the limits of what can possibly be conceived.[1] This imaginary extends to moral thinking about technology, such that whenever one writes in critique of technology—something that seems like such an unqualified good—the inevitable question asked of the critic is: "What? You're against helping people?" Technology and its potential for goodness seem like "no brainers." Its logic is so clear. The utilitarian moral assessment is a very quick. (1) Calculate the goods; maximize them; (2) Calculate the

[1] Taylor, *Modern Social Imaginaries.*

J. P. Bishop (✉)
Saint Louis University, St. Louis, MO, USA
e-mail: jeffrey.bishop@slu.edu

S. Donaldson, R. Cole-Turner (eds.), *Christian Perspectives on Transhumanism and the Church*, Palgrave Studies in the Future of Humanity and its Successors,
https://doi.org/10.1007/978-3-319-90323-1_7

harms; minimize them; (3) Make sure of the just distribution of the technology; (4) Crunch the numbers in terms of prognosticated goods and harms. Moral assessment done! On this view, technology is mere means—a neutral means—to which we supply an imminent end. Technology can be deployed for good or ill, and the majority of the time we only need worry about the unintended consequences. Medical technology is just an achievement of science, a neutral instrument deployed to enact the projected goods—or so the story goes.

The social imaginary of our time only permits questions like: "What does little Tommie want (what does he will)?"; "What do little Tommie's parents want?" Metaphysical questions, such as "What's a body for?" or "What would we have to believe about a body—or the nature of the human person—such that we could claim the technology is legitimately deployed?" go unasked. Few inquire, "What does little Tommie think his body is or what is the purpose of his body?" or rather, "What does little Tommie's parents take his body to be and what do they take the purpose of little Tommie's life to be?" Specific metaphysical or ontological questions sound odd to the modern ear.

In this essay, I shall argue that enhancement technology is not merely a scientific or technical achievement; it is not merely an achievement of the scientists and the engineers. It is not merely the achievement of the "how" of the world. That is to say, technology is not merely a means, but is itself tied up with the ends of a culture. Enhancement technology is rather a cultural and political achievement. Its fruits seem sweet; yet, I shall argue that it is the achievement of a rather dark view of the world. It is the achievement of a sinister metaphysics, originating from relatively recent Western cultural ideas about the ambiguity of the body. These ideas shape the work of contemporary Western medical science and technology, including enhancement technology. Despite the rhetoric put forward by proponents of enhancement technology that science and technology are morally neutral, that they are unencumbered by tradition, history, culture, and politics, I shall argue that the science of enhancement, and its proponents themselves, are products of a particular culture and a particular view about the nature of reality, and that that cultural lifeworld adheres to a particular metaphysics I shall claim is antithetical to a Christian understanding of the world. I do not mean to impugn the good intentions of Christians who embrace transhumanism, but to suggest that the metaphysical assumptions that animate transhumanist technology are of dubious origins.

At the same time that these technologies are products of a culture—of a particular social imaginary—they also produce the scientific imagination for other futuristic possibilities, possibilities for what bodies can be in late-modern Western culture. These enhancement technologies have consequences for ecology, economics, and human thriving, no doubt, but they also have consequences for how we understand and honor atypical bodies in the West. In short, I shall argue that enhancement technologies and the whole transhumanist lifeworld cannot be merely accepted by Christians because at the heart of these transhumanist lifeworlds is a metaphysics and an ontology that is alien to Christianity.

The Body of Medicine

The West, including Western medicine, has an ambiguous relationship to the body.[2] The body is the ground for everything; yet, the frailty of the body and our ability to manipulate it means that the body is replaceable, almost nothing in itself. This ambiguous attitude toward the body is clear in many practices of contemporary Western cultures.

On the one hand, the body gets more attention than it ever has from the medical community not only to keep the body alive, but also with the prevalent use of cosmetic dermatology, cosmetic dentistry, Botox injections, and cosmetic surgery. This body obsession extends well beyond medicine, though, as is evident in the rise of the cosmetics industry, massage studios, and gyms. In Western cultures, we are obsessed with the body's dry skin, its smelliness, its hairiness, its shape, and its frailties. The clothing with which we decorate the body, the perfumes we create to mask its smells, and the dyes we use to hide gray hair all demonstrate just how obsessed we are with the body.[3] On the other hand, the body is nothing—mere matter to be manipulated for our desires and pleasures, to be shaped into our culture's notion of beauty, to be tattooed and pierced, to be shaped and molded into what we would like it to be for us or for our peer groups. In this sense, the body is but the material which is animated by the will of the individual, a will that seeks to be unfettered by the limitations of the body. The will is the subject; the body is its object to be manipulated by that subject-will.

[2] Bishop, "Body Work and the Work of Bodies."
[3] McGill, *Death and Life.*

Thus, the dermatologist not only treats skin cancers and skin infections—frailties of the body—but she also has techniques and creams to prevent wrinkling. The dentist not only treats tooth decay and pulls bad teeth, but also whitens and straightens the teeth for cosmetic purposes. The psychologist or psychiatrist not only treats brain or psychological disorders, but also assists in optimizing one's potential through the use of self-help techniques or stimulants. If we can optimize the potential, why not maximize it, or even supersede the maximized potential?

The drive to return a failing body to a naturally functioning state—the focus of traditional medicine—eventually gives rise to a drive aimed at optimizing the function of the body. But in many instances, for example performance athletics, we seek to maximize the body's potential, taking it to the limit of natural capacities. With transhumanism, we seem to want to exceed those limitations. These achievements in medicine are not merely technical; they are cultural achievements beyond the merely scientific understanding of the body. What do I mean by a "cultural achievement?" I mean that our notions of "normal" or "natural" functioning arise from within a culture, which means that they always already carry with them a moral valence. In a culture obsessed with athletics, we see more development in nutritionals, workout regimens, massage therapy, etc. In a culture obsessed with beauty we find more spas, salons, and cosmetic surgical companies that are also spas. So, what is "natural" is cultural through and through.

The insight that science and technology are cultural achievements is not new.[4] In fact, Ernst Cassirer (1874–1944) offers the best articulation of just such a philosophy of culture and its relationship not only to science, but also to religion.[5] His mature work attempted to bridge the divide between the sciences and the humanities. He eschewed the positivism regnant in the Western science of his day, but he also shunned the irrationalism of post-Nietzschean philosophy. He claimed that both science and the humanities originated in the rational creativity of the human mind.[6]

Cassirer, billed as the last philosopher of culture, understood science to be one symbolic form among others.[7] By "symbolic form," Cassirer meant

[4] Cassirer, *An Essay on Man*; Fleck, *Genesis and Development of a Scientific Fact*; Kuhn *The Structure of Scientific Revolutions.*

[5] Cassirer, a philosopher, was a German Jew who was forced to flee Germany when the Nazis came to power.

[6] Cassirer, *An Essay on Man.*

[7] Skidelsky, *Ernst Cassirer.*

that science does not immediately—that is to say in an unmediated way—capture brute facts imposed upon the human sensory apparatus by an external reality. Instead, reality is itself mediated to the human mind through symbols, and the system of symbols are culturally derived. Language, history, art, myth, religion, and even science "are the varied threads which weave the symbolic net, the tangled web of human experience."[8] That is to say, the human animal does not merely respond to a stimulus in the way that a slug might respond to a food source or a threat; the human animal interacts with her world through the mediation of these symbolic forms. The capacity for rational engagement through these symbolic forms creates the possibility for shared cultural meanings. Put differently, humankind does not live in a merely physical universe, but in a symbolic universe. There is a difference between organic reactions to reality and human responses to reality. The reality of the world is mediated to us through the symbolic forms of myth, religion, art, language, history, and even science. Each symbolic form permits different things to come into relief for us.

Cassirer calls the human being the *animal symbolicum*.[9] For Cassirer, the mathematics that represents the behavior of subatomic particles is as much a symbolic system carrying meaning to the human mind, as are the mythic symbols of cultic practices, like the Eucharist. Cassirer places science among other symbolic forms including religion, which is the symbolic form that captures the moral dimension of human knowing. Each form has a part to play in the philosophy of culture. Science attempts to abstract from anything historical, artistic, or religious, in order to say what is true independent of history, art, or religion. But science is not the highest or the best of symbolic forms, because with science you only get one aspect of reality mediated to the mind. Science is just another symbolic form with its own ways of uncovering reality. Yet, each symbolic form has a different role to play and for different human purposes. The symbolic forms and the purposes to which they are put emerge out of a culture.

What Cassirer claims to be true of science generally would also be true of medical science and technology specifically. Medical science attempts to understand the bodies of human beings in structural and physiological terms, seemingly without reference to culture. On the merely biological side, medical science has always been focused on bodies—typically ailing

[8] Cassirer, *An Essay on Man*, 25.
[9] Ibid., 26.

bodies, bodies in need of something, bodies having deviated from normal physiological functioning. Thus, medical science and technology have a purpose built into their investigations and drive to know: that purpose is to relieve the human estate of its frailties, as Francis Bacon proclaimed.[10] At the same time as one begins to understand how to fix the failing biology, one also learns how to optimize the functions of the human body. These two factors—fixing and optimizing—are the traditional domains of medical science.

Let me give a few examples. Let's say we have a patient with a stenotic heart valve. Now we can do surgery to open a scarred heart valve; or we can give a medication to improve the efficiency of heart muscle contraction; or we can place an implantable device that promotes cardiac output. All of these options improve efficiency of blood flow to the body. Yet, that which the body needs is always instantiated in a culture and carries with it a moral valence. One does not investigate the function of the failing heart in order merely to understand the mechanism of the failing heart. Instead, one investigates the needs of the failing heart in order to fix it or to optimize its functioning for certain purposes, like increasing one's ability to play with one's grandchildren. But then one has to be in a culture in which the grandchildren themselves are not in need of the funds that are going to keep grandpa alive. Thus, fixing a biological problem or optimizing the function of the heart also enacts a cultural moral vision of the good life. One has to have the capital in order to enact that moral vision.

Another example is sildenafil; it was a drug that was used to treat severe right heart failure, especially in premature infants in the ICU. Sildenafil was used to lower right heart pressures and to treat severe pulmonary hypertension. The investigators noticed that the drug had one really interesting side effect: it caused the preemies to have erections. Thus, in a culture where sexuality takes up a good portion of our cultural energy, sildenafil becomes Viagra. Outside of a culture obsessed with sex (or rich enough to be obsessed with sex), we would overlook these side effects as nothing. A whole new disease or disorder is born because a whole new industry is created to meet a need we didn't realize we had. So in our culture a new domain of research is opened up and a host of new drugs are developed to treat what had never before been a disease.

[10] Bacon, *The New Organon*; McKenny, *To Relieve the Human Condition*; Bishop, *The Anticipatory Corpse*.

Let's look at another development: the alpha-fetoprotein test. In Britain in the late 1960s and early 1970s, there were a lot of children being born with neural tube defects. These defects can be severe or minimal. For those with the defect, it can result in impairment of neurological function below the spinal level of the defect, with paralysis and inability to control bladder or bowel elimination. The National Health Service (NHS) in Britain was very interested in saving money and there was a concerted effort to find prenatal screening tools that would allow them to detect neural tube defects in utero. They figured out that alpha-fetoprotein is a marker and began a concerted campaign to screen women for fetuses with neural tube defects. The purpose of the test was in part to save money for the social services of post-war Britain by identifying the defect before birth. Of course, the mechanism for saving money was through the termination of "defective" fetuses.[11]

All tests in medicine are not just positive or negative. Those tests only gave an alpha-fetoprotein level, but scientists had to figure out the meaning of the readings. They had to decide what the cut-offs should be between normal and abnormal levels. One woman's level could be high and her fetus might be normal, and another woman's level could be lower than the first woman's level and this fetus have the neural tube defect. One gets a Bell curve of test results. All tests in medicine have false positive and false negative rates and, prior to the abortion, one could not at that time know whether the fetus being terminated did or did not have neural tube defect. In the absence of sonography, which wasn't developed until later, the researchers were left to figure out where to draw the line. Draw the line separating normal levels from abnormal levels too far to the right and "too many" babies with neural tube defect might be born; draw the line too far to the left and normal babies might be terminated. Thus, the test itself is not neutral, it was developed for certain kinds of purposes that had moral valence, and the quest to find the test and the decision to draw lines enacted that moral vision.[12]

The tests were deployed with greater zeal in poorer neighborhoods and regions all over Britain, particularly in Northern Ireland and Wales.[13] Moreover, by a trick of statistics, in a region where incidence of neural

[11] Gagen and Bishop, "Ethics, Justification and the Prevention of Spina Bifida."

[12] Ibid.

[13] Roberts et al., "The Efficacy of a Serum Screening Service for Neural-tube Defects: the South Wales Experience."

tube defects is low, there can be a higher rate of false positives. More normal fetuses will be terminated in lower incidence areas. In fact, cases exist in which women with higher levels of alpha-fetoprotein were encouraged to abort their fetuses only to find out that their child did not have neural tube defect.[14] So why the drive to deploy this technology; why the drive to develop this technology? Political and economic expediency. Thus, the alpha-fetoprotein test carried moral, economic, and political valence and was not just the result of a new development in acultural science and technology.

Western medical technology, then, is not a neutral tool, a neutral means, but is a mechanism of culture carrying the ends desired by that culture. Medicine is not first and foremost a scientific endeavor; it is first and foremost a moral and cultural endeavor. Teasing apart the scientific or the technological dimensions from the moral is actually impossible. Medicine—first in fixing, and then optimizing function, then screening for normal function—imports a moral vision into its tests and its treatments. Thus, we extend well beyond the scientific drive to optimize, such that part of the domain of medicine begins to maximize certain features of the human body—like fat metabolism or memory enhancement, or erectile dysfunction—through the use of pharmaceuticals. It even seeks to supersede the biological limitations of the function of the human body through things like human-computer interfaces. Or it seeks to remove the beings that are limited by their bodies from existence itself. Thus, science and technology enact a certain cultural moral vision because it originates in a particular understanding of the ambiguity of the body in the West.

Futuristic Science and Transhumanism

There are several implications for the expansion of medical science and technology beyond the domain of returning a patient to normal function or optimizing normal function. In order to get at these implications, I will point to two contemporary futuristic thinkers and demonstrate several moral features that are implicit in their work. The two thinkers are Aubrey de Grey and Ray Kurzweil.[15]

[14] Campbell et al., "Ultrasound in the Diagnosis of Spina Bifida;" Gagen and Bishop, "Ethics, Justification and the Prevention of Spina Bifida."

[15] de Grey and Rae, *Ending Aging*; Kurzweil, *The Singularity is Near*.

De Grey is a biogerontologist with numerous publications in scientific journals.[16] He concludes that there is nothing in the body that says that death or even aging is inevitable. In fact, aging is not a genetically defined process of life; thus it is not in the human essence that humans should age and die. Instead, aging is a process where the accumulation of wear and tear on the body builds to a point where the survival of the organism is threatened. "The human body is a machine – a massively complex one, to be sure, but still a machine."[17] Thus, like any machine, we ought to be able to carry out preventative maintenance on it. The hurdles that exist in repairing the microstructure and functions of the biological apparatus are merely technical hurdles, which should be easy to overcome. And de Grey claims that we have a moral obligation to overcome these technical hurdles in order to prevent aging and increase the human health span.[18]

Kurzweil is a computer engineer and a successful and respected innovator in the field of computer technology.[19] Kurzweil holds that three advances in the sciences—genetics, nanotechnology, and robotics (understood primarily as artificial intelligence)—will result in a development he calls the "Singularity."[20] The "Singularity," in Kurzweil's theory, is an evolutionary idea that has been developing since the dawn of time. At each epochal stage of development from the big bang forward, matter organizes itself into more and more complex structures until intelligence is achieved with the human brain. By deploying that intelligence on the material world, the human develops science until finally the merger of human intelligence and human technology creates the conditions for the possibility of the Singularity, when the universe wakes up.[21]

[16] de Grey, "$HO_2\bullet$: The Forgotten Radical;" de Grey, "Free Radicals in Aging: Causal Complexity and Its Biomedical Implications;" de Grey, "Alzheimer's, Atherosclerosis, and Aggregates: A Role of Bacterial Degradation;" de Grey et al., "Is Human Aging Still Mysterious Enough to be Left Only to Scientists?" Khrapko, et al., "Does Premature Aging of the mtDNA Mutator Mouse Prove that mtDNA Mutations are Involved in Natural Aging?"

[17] de Grey, "Zeno's Paradox and the Faith that Technological Game-changers are Impossible," 94.

[18] de Grey, "$HO_2\bullet$: The Forgotten Radical;" de Grey, "Reason and Methods for Promoting Our Duty to Extend Healthy Life Indefinitely;" de Grey, "Zeno's Paradox and the Faith that Technological Game-changers are Impossible;" de Grey and Rae, *Ending Aging*.

[19] Kurzweil, "Reinventing Humanity: The Future of Machine-human Intelligence;" Kurzweil, "The Future of Intelligent Technology and Its Impact on Disabilities."

[20] Kurzweil, *The Singularity is Near*, 205–298.

[21] Ibid., 21.

The body for these two thinkers is replaceable in principle, on the one hand with new carbon, and the other with a more stable material, silicon. Yet, these two thinkers not only share an understanding about the ambiguity of the body—that it is both everything and nothing—but they also enact a shared moral vision based upon this understanding of the body. Each popularizes philosopher Nick Bostrom's vision of transhumanism—a scientific and technological remedy for human frailty.[22] Transhumanism's roots are also grounded in the work of Francis Bacon, the father of modern empiricism and philosophical founder of modern technoscience,[23] and transhumanist claims can come across at times as surreal and almost unbelievable in their dreams for human life, or rather post-human life.

The post-human future is not without its mythological and religious elements. Nick Bostrom, in "The Fable of the Dragon Tyrant," tells the tale of a mythical monster that requires the sacrifice of human beings. Over time, humans have understood the dragon tyrant to be invincible. Some humans struggle to resist the destruction of human beings, all of whom are eventually eaten by the dragon tyrant. Other humans become complicit in this sacrifice; those who are complicit are the religious and political leaders that thwart the morally pure drive of those who would scientifically intervene to kill the dragon tyrant. According to Bostrom—who interprets his own fable—the dragon tyrant is old age, though I would claim it is death. In his myth, Bostrom's scientists have a moral vision that will allow them to slay all the defects embodied by the dragon tyrant, allowing all to live on indefinitely.[24] Both de Grey and Kurzweil share Bostrom's mythology, and the import of their work carries the same moral valence of Bostrom's scientists, who will slay death. Both de Grey and Kurzweil hold to a very different mythological/religious form—as Cassirer would describe it—than the one to which Christians have traditionally held. Their lifeworld is not merely scientific, nor is it neutral; it is already a mythological and religious form. It is wholly moral, and wholly cultural. It is rife with Nietzsche's power ontology, as I shall show.

[22] Bostrom, *The Transhumanist FAQ v. 2.1.*

[23] Bishop, "Transhumanism, Metaphysics, and the Posthuman God."

[24] Bostrom, "The Fable of the Dragon-Tyrant."

Transhumanism's Power Ontology

Thinkers like Aubrey de Grey, Ray Kurzweil, and Nick Bostrom are claiming that the body is really nothing more than the concatenation of forces that coalesce into this thing we call a whole organism. The powers of attraction and repulsion of subatomic particles coalesce into protons, neutrons, and electrons, which in turn coalesce into atomic particles, which coalesce into molecules, which coalesce into macromolecules, which coalesce into cells, which coalesce into organisms, and so forth and so on, until you get this thinking thing, this "three-pound lump of clay" doing some very complicated things that matter does when it coalesces in just this way.[25] Thus, the body's significance is simply the way that the creative forces of the natural order shape the being in this way for the moment. Eventually these thinking bodies coalesce into communities. Thus, from subatomic particles to human communities, the most basic organizing unit is power.

Friedrich Nietzsche called the creative and chaotic forces, from which matter emerges in evermore complex structures, the Dionysian element.[26] At the animal organismal level, these forces are the instinctual drives of the animal; they are the powers that all matter possesses to reach beyond itself, to coalesce as a concatenation of powers that exist in this moment and for the moment. This concatenation of forces, this will-to-power is not agential. It is what drives the lion to kill the gazelle to eat; it is the power to procreate. It is the power to organize animals into herds for survival benefit, or the power of humans to organize themselves into complex societies. Yet, at some point this power will no longer be able to sustain itself, eventually undergoing an entropic collapse only to enter the cycle once more, the eternal recurrence of the same.[27]

For transhumanists such as John Harris[28] and Bostrom, the agential will can come to shape evolutionary history, where for Nietzsche it is really a non-agential will that ultimately ends in collapse and return to the same base power. From the former we get designer evolution. The human will-to-power turns to consciously and with agency control the evolutionary

[25] Bostrom, *The Transhumanist FAQ v. 2.1*, 6.

[26] Nietzsche, "The Birth of Tragedy," 1–116.

[27] Nietzsche *Thus Spoke Zarathustra*, 119–188; Richardson, *Nietzsche's System*, 11–65; Richardson, *Nietzsche's New Darwinism*, 12–13.

[28] Harris, *Enhancing Evolution*.

movement toward the purposes of the human will.[29] Evolutionary theory is of course itself a cultural achievement, an adequate and often accurate description and explanation of the observed facts. It is now something that comes under human agency of particularly powerful people, for particular purposes.

Transhumanist philosophy is a mythological story involving the posthuman god (whether in carbon or silicon), interpreted through the lens of medicoscientificotechnological symbolic forms originating in the Western Enlightenment, aimed at control and mastery of all that there was, and is, and is to come. Thus, the very thing that Martin Heidegger critiques as the enframing, the *gestell*,[30] is being enacted in transhumanism, the harnessing of power to achieve greater power and greater aggrandizement of a particular form of human being.

The Utilitarian Default Position

Now, I realize that many of the readers of this volume will claim that some of the very things that Nietzsche's power ontology proclaims as our transhumanist future can be found in Christianity itself. While I cannot exhaustively deal with such critiques of my position, let me take two of the most apparent points of harmony between Christian theological positions and the positions of transhumanists that I have claimed are problematic. Each of these critiques hinges on what I will call the Utilitarian Default Position. The very fact that we see similitude between some Christian themes and the transhumanist position is because our moral imaginations have been constricted by the modern enframing (i.e., Heidegger's *gestell*).[31]

First, there are those that might suggest that humans are placed in dominion over the earth, and that they should subdue the earth. Yet we must distinguish between our contemporary understanding of dominion and the ancient understanding of dominion. As Martin Heidegger notes in *The Question Concerning Technology*, one can distinguish between the "bringing forth" and "challenging forth" of creation.[32]

[29] Bishop, "Transhumanism, Metaphysics, and the Posthuman God."

[30] Heidegger, "The Question Concerning Technology," in Heidegger, *Basic Writings*.

[31] Ibid., 324–325.

[32] Ibid.; Borgmann, *Technology and the Character of Contemporary Life*; Verbeek, *Moralizing Technology*.

A few analogies are in order here: (1) A midwife brings forth while an obstetrician challenges forth. That is not at all to say that when an obstetrician is doing a Cesarean section that she is doing evil. The problem arises when the technological tools and skills become the lens through which all pregnancies are viewed. (2) A farmer who cultivates the land in rotation of crops, in the tilling of the soil, growing crops in harmony with the seasons is doing something very different than the industrial farmer that pumps the land with fertilizers, waters the arid land in the dry season, and produces his crops in whatever season he wills. (3) A miller who places his paddlewheel into the stream in order to turn the mill is very different from the engineer who runs the entire river through the hydroelectric plant. The midwife, the farmer, and the miller bring forth a child, a crop, and ground corn; the obstetrician, the industrial farmer, and the engineer challenge forth the child, the crop, and the power of the river.

Because the late-modern West is fully steeped in the modern *gestell*, it is difficult for us to see the distinction between bringing forth and challenging forth. The contemporary social imaginary of the technological West is already shaped by the enframing, and the utilitarian calculus is itself part of the fabric of the enframing. We are concerned with the ends—the development of the product—no matter the means. Thus, it appears to us that there is no difference between a farmer in scripture and the industrial farmer. When God gives humankind dominion over the earth according to the agrarian authors of the Bible, I suspect they had something in mind closer to the bringing forth understood in agrarian societies rather than challenging forth of technological societies. In fact, Hebrew Scripture scholar Ellen Davis makes just this point, that the human relationship to creation is not one that lords it over the land, but lovingly works alongside the land.[33]

Second, while some of the goals of transhumanists articulated above sound very much like Christian goals, there is again a difference. Let us take as an example the desire espoused by Bostrom: "to exercise control over their own desires, moods, and mental states; to be able to avoid feeling tired, hateful, or irritated about petty things."[34] Again, to the contemporary mind, the goal or object or good to be produced holds the prime position. It seems to me that the practices and habits cultivated by a self-reflective human being—a virtue ethics approach—is very different from brain implants that might regulate the relationship between the amygdala and the cingulate gyrus.

[33] Davis, *Scripture, Culture, and Agriculture*.

[34] Bostrom, *The Transhumanist FAQ v. 2.1*, 5.

For example, as Howard Clark Kee and Gary Ferngren have shown, the goal of healing in the early church is not as important as faithfulness as one strives to achieve the goal of health.[35] As Kee notes, "What is sought [in magic] is not to learn the will of the deity, but to shape the deity's will…" by asserting one's own will over that of God.[36] Calling on the names of false gods is of particular concern to the Christian precisely because the means to achieving the ends is part of what defines the work as good, not merely the product of the work. We can find the closeness of means and ends even in language. There is no distinction between curing and caring for the ancient; both of our terms originate from the same Latin word, *curare*, to care.[37] The will, the means, and the ends had to align with the will of God.

In other words, for the early Christian, the underlying worldview, the underlying metaphysics, mattered; they did not focus solely on the effects brought about by the manipulation of power. Thus, it is entirely possible that the effect desired is a good, but that the means to achieve that good were not permitted to ancient Christians. While such a distinction is foreign to us, formed as we are within the modern technological *gestell*, it was not foreign to them. Thus, it is entirely possible that the technological challenging forth that is part and parcel of the modern enframing, where the ontology is nothing more than a power ontology, is highly problematic. That is not to say that Christianity would in all cases be against the use of technology (e.g., as espoused above by Bostrom[38]). It is to say that there might be some instances or some situations where the technological challenging forth is problematic for Christians.

The problem with both of these ways of assessing technology and its uses within transhumanism is that the utilitarian calculus is already at one with the power ontology of transhumanism. In fact, the whole way that many Christian transhumanists read scripture is already shaped by our contemporary ontotheology and the moral imagination that it produces, namely utilitarianism. It assumes that the products or the objects to be produced—the goods to be produced—are the sole concern for the Christian of any era. I am concerned that Christian transhumanists have

[35] Kee, *Medicine, Miracle and Magic in New Testament Times*; Ferngren, *Medicine and Health Care in Early Christianity*.

[36] Kee, *Medicine, Miracle and Magic in New Testament Times*, 112.

[37] Ferngren, *Medicine and Health Care in Early Christianity*, 127.

[38] Cf., Bostrom, *The Transhumanist FAQ v. 2.1*, 5.

already accepted the world to be little more than the concatenation of forces described in Nietzsche's power ontology and subsumed under the banner of transhumanism. They have dressed up their language in the language of co-creation. Transhumanism already presumes the world to be a certain way such that the utilitarian calculus—with its concern for products—is the only mode of moral analysis with which contemporary humankind needs to concern himself. But the question that seems to slip beneath the modern threshold is this: Is the world really what Nietzsche's power ontology says it is? Or is it what our modern moral imaginations have made it?

For the Life of the World

Is the world in reality a concatenation of forces, originating in the big bang and resulting in this thinking thing that sits before you with its Luddite tendencies? Some will call the concatenation of forces "deep evolution." I hold to some aspects of the description of those facts, but I interpret those facts very differently. First, I hold that their emergence as facts are part of a cultural process sometimes called the scientific method. Or as Francis Bacon puts it, nature must be harassed to reveal her secrets.[39] Put differently, "technology is a mode of revealing".[40]

Second, if we accept Cassirer's and even Heidegger's rendering that we are the *animal symbolicum*, that we are the being that is always already in the world, thrown not merely into space and time, but into cultural spaces and historical times, then it is not that the world is thus, but rather it is thus that we have made the world through our symbolic forms, our ways of seeing and apprehending the world, and our ways of being in the world. I would argue, then, that the claim that the transhumanist way of being in the world—even the Christian transhumanist way of being in the world—is the only symbolic mediation that is true is highly suspect. To read the world and scripture through this lens, it seems to me, is to do violence to the world and to scripture. After all, if the world is power all the way down, then could we expect the world to be anything other than violent; does it not lead us to believe that the world is nothing more than solitary, poor, nasty, brutish, and short, and that it is red in tooth and claw? This

[39] Bacon, *The New Organon*, 20.

[40] Heidegger, "The Question Concerning Technology," in Heidegger, *Basic Writings*, 319.

violence begins in Bacon's science and ends in the hydroelectric plant, which takes nature and runs it through its machinery for human purposes, destroying the river as it was, and the surrounding land.

For Christians, I would say that our experience of the resurrected Christ results in a different symbolic form; the resurrected body is our symbolic form, par excellence. The Eucharist mediates to us a different way of reading nature, such that nature is not nature, but it is instead creation, redeemed creation.[41] We co-create not just by making things out of the stuff of the world, but we co-create the way the world appears to us and for us. If what Christians hold about creation is true, if what Christians hold about creation and the redemption of creation is true, then it is most certainly the case that the transhumanist interpretation of nature as the concatenation of forces redirected toward the next iteration of the *Übermensch*, the post-human, is nothing more than idolatry. It seems that the transhumanist proclaims the world to be in the image of power and that we humans can turn our powerful intellect onto that world to make it in whatever image we concoct. At the heart of all forms of humanism—even transhumanism—sits the Protagorean proclamation: the human mind is the measure of all things. And the thing about our humanist construction of all things—that is to say, the thing about the humanist idolatries—is that humanisms are very quick to sacrifice those beings that don't measure up to the imagined gods. We must remember, as uncomfortable as it makes us feel, that National Socialism, Soviet Communism, and Capitalism are all humanisms. On the Christian rendering, differently embodied creatures—the sick, the lame, the deaf, the blind—are welcomed and are not defective arrangements of material forces. They are fellow creatures deserving of their own dignity even if, perhaps especially if, they don't measure up. They are welcomed as they are. Even if they cannot be cured or made to fit the normalizing gazes created by the medicoscientificotechnological symbolic form of transhumanism, they are welcomed as they are.

So, if what we Christians hold about the world is true, then creation lends itself to a myriad of symbolic forms; it lends itself to a myriad of ways of being-in-the-world. Creation is diverse, in part because our lifeworlds and our symbolic forms are diverse. Thus beings like those with heart failure and spina bifida are capable not only of surviving, but also thriving, even in their frailty if we only see them rightly, if we only engage them

[41] Wirzba, *From Nature to Creation*.

rightly. Thus, war of each against each, of survival of the fittest, of "natural" selection, and even so-called designer evolution[42] are all ways of interpreting the realities of creation. But the Christian way is to speak about a graced world where the poor are selected not to be manipulated, but such that we assist them in thriving; where the weak are selected because existence is itself a good of creation, just as God exists; where we preferentially select those that transhumanist society says ought to be selected out (even if they only say it implicitly). We choose to bring them to our banquet tables because there may be no way to fix them and it is not even clear that they think they need fixing.

The Christian message of resurrection is that bodies matter, they have significance, and they are not just clay to be molded to our wills. The cruciform and resurrected Christ is our symbolic form. He is our way of being-in-the-world, not a will over the world. The world lends itself to this mediation because of the Eucharist where the really real comes into being for us, not in the transhumanist transubstantiation of ourselves into what we will, but in the word made flesh for us in the Eucharistic meal.

For Christians, that means we are not what the scientists and technologists say we are. We are not the concatenation of forces, some creative and some destructive, but creation is itself graced—these frail bodies that we *are* are themselves graced, even in their frailties and their dependencies. Our participation in the being of God reveals that we are not in a state of nature of each warring against each, red in tooth and claw, of outwitting each other so that the capitalist framework necessary to sustain the kind of quest for the post-human state destroys the very thing it purports to enable. The resurrected body of Christ is not Kurzweil's silicon body of the singularity. The resurrected body of Christ is confounding to all humanisms, even the really smart transhumanist men and women who would wield power over those who lack it. The resurrected body of Christ will appear as folly to the power ontology of transhumanism. The adolescent longing to have superhuman strength and superhuman intelligence, the desire to outwit all that have gone before us—the desire to be like God—is just a different version of the survival of the fittest, a different kind of war where some will be left out. The god of that mythology is the post-human idol, to which we as Christians cannot sacrifice and to which we cannot bend the knee.

[42] Young, *Designer Evolution*.

Bibliography

Bacon, Francis. 2000. In *The New Organon*, ed. Lisa Jardine and Michael Silverthorne. Cambridge: Cambridge University Press.

Bishop, J.P. 2010. Transhumanism, Metaphysics, and the Posthuman God. *Journal of Medicine and Philosophy* 35 (6): 700–720.

———. 2011. *The Anticipatory Corpse: Medicine, Power and the Care of the Dying*. South Bend: University of Notre Dame Press.

———. 2013. Body Work and the Work of Bodies. *Journal of Moral Theology* 2 (1): 113–131.

Borgmann, Albert. 1987. *Technology and the Character of Contemporary Life*. Chicago: University Of Chicago Press.

Bostrom, Nick. 2003. *The Transhumanist FAQ v. 2.1*. Oxford: World Transhumanist Association.

———. 2005. The Fable of the Dragon-Tyrant. *Journal of Medical Ethics* 31 (5): 273–277.

Campbell, S., J. Pryse-Davies, T.M. Coltart, Mary J. Seller, and Jack D. Singer. 1975. Ultrasound in the Diagnosis of Spina Bifida. *Lancet* 305 (7915): 1065–1068.

Cassirer, Ernst. 1992. *An Essay on Man*. New Haven: Yale University Press.

Davis, Ellen. 2009. *Scripture, Culture, and Agriculture: An Agrarian Reading of the Bible*. Cambridge: Cambridge University Press.

de Grey, Aubrey D.N.J. 2002. HO_2•: The Forgotten Radical. *DNA and Cell Biology* 21 (4): 251–257.

———. 2006. Free Radicals in Aging: Causal Complexity and Its Biomedical Implications. *Free Radical Research* 40 (12): 1244–1249.

———. 2007. Alzheimer's, Atherosclerosis, and Aggregates: A Role of Bacterial Degradation. *Nutrition Reviews* 65 (12): S221–S227.

———. 2008. Reason and Methods for Promoting Our Duty to Extend Healthy Life Indefinitely. *Journal of Evolution and Technology* 18 (1): 1–6.

———. 2013. Zeno's Paradox and the Faith that Technological Game-Changers Are Impossible. *Gerontology* 59: 93–94.

de Grey, Aubrey D.N.J., and M. Rae. 2008. *Ending Aging: The Rejuvenation Breakthroughs that Could Reverse Human Aging in Our Lifetime*. New York: St. Martin's Griffin.

de Grey, Aubrey D.N.J., Jown W. Baynes, David Berd, Christopher B. Heward, Graham Pawelec, and Gregory Stock. 2002. Is Human Aging Still Mysterious Enough to be Left Only to Scientists? *BioEssays* 24 (7): 667–676.

Ferngren, Gary B. 2009. *Medicine and Health Care in Early Christianity*. Baltimore: Johns Hopkins University Press.

Fleck, Ludwig. 1979. *Genesis and Development of a Scientific Fact*, ed. Thaddeus J. Treen and Robert K. Merton, Trans. Fred Bradley and Thaddeus J. Trenn. Chicago: University of Chicago press.

Gagen, Wendy Jane, and Jeffrey P. Bishop. 2007. Ethics, Justification and the Prevention of Spina Bifida. *Journal of Medical Ethics* 33 (9): 501–507.

Harris, John. 2007. *Enhancing Evolution*. Princeton: Princeton University Press.

Heidegger, Martin. 1977. In *Basic Writings*, ed. David Farrell Krell. San Francisco: HarperCollins.

Kee, Howard Clark. 1988. *Medicine, Miracle and Magic in New Testament Times, Society for New Testament Studies Monograph Series*. 1st ed. Cambridge: Cambridge University Press.

Khrapko, Konstantin, Yevgenya Kraytsberg, Aubrey D.N.J. de Grey, Jan Vijg, and Eric A. Schon. 2006. Does Premature Aging of the mtDNA Mutator Mouse Prove that mtDNA Mutations Are Involved in Natural Aging? *Aging Cell* 5 (3): 279–282.

Kuhn, Thomas. 2012. *The Structure of Scientific Revolutions*. 4th ed. Chicago: University of Chicago Press.

Kurzweil, Ray. 2003. The Future of Intelligent Technology and Its Impact on Disabilities. *Journal of Visual Impairment and Blindness* 97 (10): 582–584.

———. 2006a. *The Singularity Is Near: When Humans Transcend Biology*. New York: Penguin.

———. 2006b. Reinventing Humanity: The Future of Machine-Human Intelligence. *The Futurist* 40: 39–46.

McGill, Arthur Chute. 1987. *Death and Life*. Philadelphia: Fortress Press.

McKenny, Gerald P. 1997. *To Relieve the Human Condition: Bioethics, Technology, and the Body*. Albany: State University of New York Press.

Nietzsche, Friedrich. 1999. The Birth of Tragedy. In *The Birth of Tragedy and Other Writings*, ed. Raymond Geuss and Ronald Speirs. Cambridge: Cambridge University Press.

———. 2006. *Thus Spoke Zarathustra*. Cambridge: Cambridge University Press.

Richardson, John. 1996. *Nietzsche's System*. New York: Oxford University Press.

———. 2004. *Nietzsche's New Darwinism*. New York: Oxford University Press.

Roberts, C.J., B.M. Hibbard, G.H. Elder, K.T. Evans, K.M. Laurence, A. Roberts, J.S. Woodhead, I.B. Robertson, and M. Hoole. 1983. The Efficacy of a Serum Screening Service for Neural-Tube Defects: The South Wales Experience. *Lancet* 1 (8337): 1315–1318.

Skidelsky, Edward. 2009. *Ernst Cassirer*. Princeton: Princeton University Press.

Taylor, Charles. 2004. *Modern Social Imaginaries*. Durham: Duke University Press.

Verbeek, Peter-Paul. 2011. *Moralizing Technology: Understanding and Designing the Morality of Things*. Chicago: University of Chicago Press.

Wirzba, Norman. 2015. *From Nature to Creation: A Christian Vision for Understanding and Loving Our World*. Grand Rapids: Baker Academic.

Young, Simon. 2006. *Designer Evolution: A Transhumanist Manifesto*. Amherst: Prometheus Books.

CHAPTER 8

The Failed Fictions of Transhumanism

Christina Bieber Lake

Introduction

In Greek mythology, when the gods fall in love with humans it usually ends badly. One of the more gruesome of such tales is that of Tithonius. Unfortunately for Tithonius, Eos, goddess of the dawn, fell in love with him and wanted him to live forever. In tears she appealed to Zeus to grant Tithonius immortality. Eos, however, forgot to ask Zeus to give her lover eternal youth. So Tithonius was trapped forever in a decaying body, eventually losing all mobility. Eos took pity on him before she left him for a younger lover. She changed Tithonius into a cicada, where he eternally buzzes out his one request: please let me die.

According to organizations like Humanity+, attaining immortality is the highest priority of the transhumanist movement. Since most transhumanists also believe that they are smarter than Eos, they always add the aspiration to be "better than well." Transhumanism promises to either reverse aging, help us to achieve technologically enhanced bodies, or provide us with completely new bodies as needed. Any opposition to this goal is seen as shortsighted, for one day everyone will want

C. B. Lake (✉)
Wheaton College, Wheaton, IL, USA
e-mail: christina.lake@wheaton.edu

S. Donaldson, R. Cole-Turner (eds.), *Christian Perspectives on Transhumanism and the Church*, Palgrave Studies in the Future of Humanity and its Successors,
https://doi.org/10.1007/978-3-319-90323-1_8

what transhumanism is promising. John G. Messerly, an outspoken transhumanist, insists that "technologically guaranteed immortality will end most people's opposition ... when immortality is real, most will choose it rather than dying and hoping for a heavenly reward."[1]

But there is more than one way to read a Greek myth. To interpret the moral of Tithonius's tale as "be sure to cover your bases when you ask for stuff" is a simplification to say the least. Instead, the Greeks might have been warning us to ask a simple question before we move forward with our desires: Do we really know what we are asking for? And, as the existence of such myths suggests, perhaps the Greeks are warning us that we should use our moral imaginations to explore what the real costs of our desires might be. But transhumanism itself has been so overtaken by techno-optimism that its moral imagination has been stunted. Full of the shimmering promise of the vision, it is blinded to the ways the vision itself might be misguided. So when proponents of this vision try to tell stories undergirded by it, all its flaws become fully apparent. What transhumanist fiction reveals is that the attitudes and values that motivate the movement are inconsistent with what is necessary to attain the perfect outcome it desires and promises.

No Place: Utopian Visions

Transhumanists are not the first folks to attempt to write a story that imagines an ideal future. Thomas More is credited with writing the first utopian fiction in 1516, but his book is not what most people think of when they think of utopia. More's book is a satire on such idealistic visions. In fact, he coined the word "utopia": it means "no place."[2] Novels that earnestly represent a perfect society did not exist until the reform-minded nineteenth century, which produced Edward Bellamy's *Looking Backward* and William Morris's *News from Nowhere*, both socialist fantasies, and both almost unreadable. Utopian novels took a new direction in the twentieth century, as in the behaviorist dream of B.F. Skinner's *Walden II*, and the nebulous reification of "Good Being" that comprises Aldous Huxley's

[1] Messerly, "Jaron Lanier on Transhumanism."

[2] Margaret Atwood writes that "perhaps he meant to indicate that although his Utopia made more rational sense than the England of his day, it was unlikely to be found anywhere outside a book." (Atwood, *Negotiating with the Dead*, 93).

Island.[3] These books are dissatisfying primarily because the citizens populating them are barely a step away from cardboard cutouts, and none of the utopias provide blueprints that sane people would actually choose to pattern an ideal society around.

The problem with all efforts to imagine a perfect world is that they usually require writers to imagine perfect future people, and this, as Charles Rubin has argued, is incomprehensible.[4] This partly explains why, although there are many visions of the ideal transhumanist future, none of them has been fully fleshed out in fiction.[5] Consider one of the founders of Humanity+: Nick Bostrom. When it comes to illustrating why it would necessarily be better to have these greater capacities, his descriptions remain abstract and theoretical. The closest thing he offers to an image of the future is his "Letter from Utopia" written by "your possible future self." The letter reads: "[W]hat you had in your best moment is not close to what I have now – a beckoning scintilla at most. If the distance between base and apex for you is eight kilometers, then to reach my dwellings would take a million light-year ascent. The altitude is outside moon and planets and all the stars your eyes can see. Beyond dreams. Beyond imagination. My consciousness is wide and deep, my life long."[6]

[3] My characterization of "earnest" utopian fiction distinguishes these idealistic texts from the "critical utopias" that flourished in the mid to late twentieth century. As Tom Moylan argues, these novels, including a number of feminist utopias such as *The Female Man* by Joanna Russ and *Woman on the Edge of Time* by Marge Piercy preserved the "subversive imaging of utopian society and the radical negativity of dystopian perception." Utopian writing was thereby "saved by its own destruction and transformation into the 'critical utopia.'" (Moylan, *Demand the Impossible*, 10).

[4] "By defining human dignity in terms of ceaseless self-overcoming, the transhumanists open the door to an incomprehensible human future." Charles Rubin, "Human Dignity and the Future of Man," 160.

[5] The closest thing to it is the Culture series by Iain Banks and *Glasshouse* by Charles Stross, neither of which actually inhabits a future enhanced-to-perfect world, and so cannot be called utopian. The Culture series imagines a "post-scarcity" world run by artificial intelligence (AI). In it, the AI are chosen to be leaders precisely because they are "good"—defined, notably, as not "bad," that is, they cannot be corrupted. The society has supposedly conquered suffering, death, and other ills. But when a society is perfect, nothing can really happen, so all the conflict comes from the Culture's interactions with the outside world, and the "perfect" world is usually only referenced, not visited. Stross's *Glasshouse* references a similar world; people born in 2050 can live forever by downloading their consciousness into new bodies. But to generate the conflict necessary for a novel, Stross sets all the action in a "glasshouse experiment" in which people from this future world are transported into a world that resembles the "dark ages." In other words, they are put in a world that looks like ours, so that things can begin to happen.

[6] Bostrom, "Letter from Utopia," 1–7.

The rest of the letter returns to explaining how to achieve that perfect self, which amounts to using technology to defeat mortality, suffering, and cognitive limitations.

Other proponents of transhumanism are equally abstract. One of the principal members of Humanity+, Aubrey de Grey, has devoted his life to bioregeneration research, the goal of which is to end physiological aging. Although admittedly there is no reason for medical science not to treat cellular aging like any other disease, or even to defeat aging as he defines it, the surrounding utopian rhetoric promises much more than these goals suggest. De Grey dedicates his book to "the tens of millions of people whose indefinite escape from aging depends upon our actions today," and refers to aging *only* as a disease, without any real engagement with substantive metaphysical objections (such as those raised by Gilbert Meilaender and many others) to doing so.[7] In short, very few thinkers associated with transhumanism seem to question the main assumption that people who live longer with younger cells will necessarily live happier or more rewarding lives. To the objection that the transhumanist vision remains unsatisfactorily abstract, Bostrom and others respond by arguing that this future world will be so different from what we know that we cannot comprehend or effectively illustrate it.

Because this utopian vision cannot be fully imagined, any transhumanist who wants to write a novel must either set it in the present or in the not-too-distant future. One such novel is *The Transhumanist Wager*, by the world-travelling reporter Zoltan Istvan.[8] Istvan was the Transhumanist Party's candidate for president in 2016, marking him as one of transhumanism's most public faces. Because *The Transhumanist Wager* has many flaws that have nothing to do with transhumanism, some will accuse me of cherry-picking.[9] As I mentioned above, it is difficult to find fully imagined transhumanist utopias to compare with Istvan's, so I had to settle on letting this novel exemplify the types of problems that I believe would face any fiction writer with a utopian (and particularly a transhumanist utopian) agenda. Although anyone can write a bad novel for reasons unrelated to

[7] De Grey and Rae, *Ending Aging*. For a response, see Gilbert Meilaender "Thinking About Aging."

[8] Istvan, *The Transhumanist Wager*.

[9] For example, there is quite a bit of poor writing. In a scene designed to depict what happens to a preacher after the woman he is lusting for mentions her husband, Istvan writes that "the courtship dance halted for Belinas. The word 'husband' was like a dagger in his cranium. He felt the sexual edge in him rescind, the fuel in his groin ooze away, the world around him instantly deflate" (Ibid., 143).

the novelist's grounding worldview, I contend that this novel's most prominent flaws are the ones that specifically reveal the failed moral imagination of transhumanism. *The Transhumanist Wager* reveals a thin understanding of character and conflict, a faulty vision of the good life, and a pervasive contempt for human persons. The novel's failures uncover some ugly truths underneath the techno-human quest for immortality.

A Thin Understanding of Character and Conflict

The novel as an art form would not have gotten anywhere without the idea of persons who are in conflict with themselves, their situations, and other persons. The Greek word "pro" from which we get the word protagonist does not mean "proponent of a cause," but "first." So protagonist means "first actor," and it initially referred to the main character in a play, the primary person who is doing something. As protagonists have developed through ancient tragedy, renaissance drama, and into the birth of the novel, we have watched them struggle to make their way through the world, to face obstacles, to grow, to change. Their frustrations are our frustrations. The better novels are the ones that contain the richest characters, characters full of recognizable flaws who find themselves in recognizable situations, even when those situations are set in the future or on different planets or different worlds.

Istvan, like transhumanism itself, is not interested in real people with these kinds of flaws and these kinds of conflicts. Transhumanism is driven by an ideological agenda to sell us a future world where people become better than themselves, live forever, and escape this kind of conflict altogether. Its main goal as a movement is to persuade us to believe that technology has the means to make us smarter, more beautiful, and ageless, and that this in turn will solve all of the petty problems we have with others. Simon Young is an example of a transhumanist thinker who, like Istvan, just assumes that if we work toward this kind of perfection, we will be better people. In *Designer Evolution* he argues that, "if we are all neurochemically empowered, perhaps we may really be able to start 'loving thy neighbor as thyself.' As we become able to create the chemistry of enlightenment through biotechnology, perhaps we may all come to feel the mystic's bliss of universal love."[10] The work of human perfection will be completed by technology.

[10] Young, *Designer Evolution*, 246.

That this is a lot to ask for from neurochemical enhancements is made clear from the protagonist of *The Transhumanist Wager*, Jethro Knights. Jethro is a thinly veiled version of Zoltan Istvan, with an equally flimsy plan to perfect human nature. Unsurprisingly, Jethro does not live after the promised enhancements have arrived, because that would force the novelist to give flesh to perfect people and then have those perfect people not have any petty conflicts. Since that is literally unimaginable (not to mention, boring), Istvan has no choice but to position his novel in the present, and to make his character more of a proponent of ideas than a protagonist. Jethro is a two-dimensional mouthpiece for transhumanism, and the things that happen to him are all related to the black-and-white universe that Istvan has set up: there are the transhumanists (good) and anyone who might be trying to stop transhumanists (bad). Most often, this latter group consists of irrational Christians who "just don't understand" what they are fighting against.

Jethro is not only flat as a character, he is also actually completely unlikeable, and unlikeable in a way that Istvan doesn't even seem to recognize. One of the worst sins a novelist can commit is not in making a character behave badly, but in making him into someone that readers cannot identify with in any way. At the beginning of the novel Istvan lists Jethro's guiding principles, unwittingly stacking the deck against our liking (or even recognizing) his hero:

1. *A transhumanist must safeguard one's own existence above all else.*
2. *A transhumanist must strive to achieve omnipotence as expediently as possible—so long as one's actions do not conflict with the First Law.*
3. *A transhumanist must safeguard value in the universe—so long as one's actions do not conflict with the First and Second Laws.*[11]

Jethro turns out to be as self-serving as his laws suggest. In spite of this, he is still able to get the girl, influence people for the cause, and dominate global politics. At no point does his self-serving nature cause him any long-term problems. Instead, all of Jethro's problems are caused by outside antagonists, as if pursuing these three laws would be workable for everyone as long as there is universal agreement with the laws. There's no meaningful conflict that builds empathy for us to follow, and the novel is reduced to ineffective propaganda.

[11] Istvan, *The Transhumanist Wager*, 4.

A Faulty Vision of the Good Life

As I mentioned above, anyone with any worldview can write propaganda that masquerades as a novel. But transhumanists who believe that achieving the stated goals of longevity, superintelligence, and super well-being will *inevitably* improve life on earth must work to hide any faulty links between these goals and the good life they presumably lead to. In other words, one can never question the assumption that these three things will *necessarily* lead to a good life. This is one reason why a transhumanist can't risk telling a real story that encapsulates their vision. Characters placed in scenarios in which these goals have been realized (utopias) will immediately reveal this assumption to be false. The characters are either going to be idealized, two-dimensional characters or they are going to find out quickly that their long lives and newfound abilities did not automatically solve their problems.

To see how the transhumanist vision is inconsistent with the good life, consider one of the stated goals of the movement: "super well-being." This goal has been used to support everything from germ-line genetic engineering, enhancement drugs, cosmetic surgery, and gene doping. Transhumanists correctly assume that no one would disagree with the general goal of alleviating people's suffering. Indeed, all world religions have been committed to this cause. But the goal of "super well-being" is not just about alleviating actual suffering. Instead it casts a very large net for what it defines as suffering—basically, anything that is an impediment to our happiness and freedom—and then assumes that all such "suffering" is bad.

When seen this way, it is clear that what underlies the false assumption that these enhancements will lead to the good life is a very thin understanding of suffering. Transhumanism not only defines suffering broadly, it also assumes that the removal of all suffering will be inherently good for human beings. But this assumption flies in the face of thousands of years of theological and philosophical argument that insists that though suffering should never and need never be sought after, it is essential for us to experience in order to grow in character and to value the lives we actually have. For example, in her book *Love's Knowledge*, Martha Nussbaum explains that there are different kinds of transcendence that humans can strive for. The first is an appropriate kind of internal transcendence: to be the best human persons we can be with the knowledge that what we are is limited in some way by what a human person is *meant* to be. That idea is, of course, Aristotelian in conception. Human beings have been created with some appropriate *telos* available to them. Christianity teaches something very similar: the fact of our being creatures and not the creator provides us with the path to an appropriate kind of striving, which is striving to be like Jesus, longing for closeness with God. But the other kind of transcendence, the wrong kind, is a desire to move beyond being human and achieving a godlike escape from death and suffering. Nussbaum argues that the ancients had a name for this: hubris.

> There is a kind of striving that is appropriate to a human life, and there is a kind of striving that consists in trying to depart from that life to another life. This is what *hubris* is—the failure to comprehend what sort of life one has actually got, the failure to live within its limits (which are also possibilities), the failure, being mortal, to think mortal thoughts. Correctly understood, the injunction to avoid *hubris* is not a penance or denial—it is an instruction as to where the valuable things *for us* are to be found.[12]

Nussbaum is not arguing that we should embrace suffering. Neither is she arguing that we should stop trying to alleviate aging, or even to stop striving to live longer and better lives. She is arguing that when the desire for transcendence crosses the line into a transcendence that makes human life incoherent, it becomes an impediment to discovering the real value of our lives. In other words, improper, hubristic striving blocks our ability to see what the good life looks like for us right now.

In *Truth and Method*, Hans-Georg Gadamer makes a similar point in his account of human experience. He also draws upon the wisdom of the Greek tragedians, particularly Aeschylus. Gadamer explains that

> What a man has to learn through suffering is not this or that particular thing, but insight into the limitations of humanity, into the absoluteness of the barrier that separates man from the divine. It is ultimately a religious insight—the kind of insight that gave birth to Greek tragedy.
>
> Thus experience is experience of human finitude. The truly experienced person is one who has taken this to heart, who knows that he is master neither of time nor the future. The experienced man knows that all foresight is limited and all plans uncertain.[13]

[12] Nussbaum, *Love's Knowledge*, 381.
[13] Gadamer, *Truth and Method*, 365.

Everything that Gadamer argues that we must learn from experience comprises concepts that are rejected by transhumanism at the outset: the limitations of humanity, the difference between creatures and the creator, and our inability to control the future. Without learning the lessons of finitude and contingency, we set ourselves up for perpetual dissatisfaction.

It is no surprise that both Nussbaum and Gadamer foreground human experience in the study of philosophy. This also explains why both of these philosophers have been so useful to literary studies. The idea that we learn primarily through experience best explains why we read and write fiction—particularly novels. *The Transhumanist Wager* fails as a novel at the outset because its author agrees with Jethro's assumption that experience should not be a teacher. Instead, any difficulties we encounter on the way to the end goal are just part of the problem; the only real struggle is to stay alive as long as we can. Needless to say, this does nothing to help us learn how to live the good life today. The good life is always to be realized tomorrow—after we make way for all the necessary technological enhancements.

Contempt for Persons

To argue that there is an appropriate kind of transcendence that is different from a hubristic kind does not mean it is easy to discern when one becomes the other. Nussbaum therefore rightly asks: "when does the aspiration to internal transcendence become the aspiration to depart from human life altogether?" and argues that there can and should be no clear answer. But I respectfully disagree with this response. I believe that the ways in which *The Transhumanist Wager* fails as a novel show us the difference. Within transhumanism (and indeed within in any reform movement), aspiration is revealed to be hubris when it comes out on the "yes" side of one question: Does this movement emerge from or produce contempt for any actual persons who are alive today?

One of the clearest insights into this question can be found in Wendell Berry's essay "Standing by Words." In transhumanism, all energy is directed toward the envisaged utopian future. Their political platform is largely defined by a shift of priorities away from defense, social security, and welfare, and toward technology that will give (some of us!) the good life free from suffering. But Berry helps us to see that this can never be love for real, other persons. It is love for an abstract future. And what is worse, it is only the basest kind of self-love. Berry explains that "one cannot love the future or anything in it, for nothing is known there. And one cannot unselfishly make a future for someone else. Love for the future is self-love—love for the present self, projected and magnified into the future, and it is an irremediable loneliness."[14] *The Transhumanist Wager* makes no effort to hide that this most base form of self-love is at its very center. In fact, it celebrates it. Jethro deliberately inverts Jesus's teaching that if you love your life in this world you will lose it. His unpoetic substitute reads that "*if you love life, you will always strive to reach the most advanced form of yourself possible while protecting that life.*"[15]

Not surprisingly, loving your own life and protecting your own future produces contempt for anyone and anything that stands in the way of that. *The Transhumanist Wager* does not try to hide that contempt. It features many speeches like this one, which Jethro delivers to a politician who is not on his side:

> What prompts alarm in me is how you and your government want to ruin not only the potential of this country, but also the path of those who are going to transition into more advanced beings in search of immortality and omnipotence, and maybe even participate in a great singularity. These advances are going to pass, one way or another. And your current second-rate moral system—your weak, pretend-God-will-take-care-of-us bullshit—is a waste for our species' possibilities. You people want to pretend that democracy, religious inspiration, and unbridled consumerism are going to last forever and carry us all to bliss; that the American Dream is right around the next corner for everyone. You spend hundreds of billions of dollars on lazy welfare recipients, on mentally challenged people, on uneducated repeat criminals, on obese second-rate citizens bankrupting our medical system, on murderous war machines fighting for oil and your oligarchy's pet projects in far off places.[16]

Jethro's contempt for the weak or underprivileged proves to be no different from Hitler's. He repeatedly calls the current culture "baggage culture" because it is weighing down the enlightened. When Transhumania (Jethro's floating utopia) invites the lowlifes from the rest of the planet to apply for help from their superior technology, we overhear the "Immortality Grant Team" discussing who is qualified to get this help. A member of the

[14] Berry, *Standing by Words*, 61.
[15] Istvan, *The Transhumanist Wager*, 179.
[16] Ibid., 127.

team is sorting through applications and tells Jethro that an unemployed woman with cancer who has applied says she "wants to live longer so she can teach her kids how to be responsible, upright people." A computer engineer asks: "[I]sn't there a way to screen idiots like that from the applicant pool? What a waste of our time. Send her six feet of rope to hang herself." To which Jethro responds, "Negative ... the cost of the rope isn't worth it."[17] It is therefore not surprising that after Jethro gains control of the world's resources through Transhumania, he authorizes the genocide of large groups of "unworthy" people. It is the logical end of the transhumanist vision for humanity's perfection—what Flannery O'Connor referred to as the tenderness that leads to the gas chamber.[18]

Charles Rubin explains that although the transhumanist conception of human dignity might look like love for self and others, it is actually self-conscious negation. It is not an affirmation of our dignity that being created in the image of God provides. It is contempt for one's own nature that is immediately followed by contempt for others. "What does it mean to say that our dignity resides in the fact that by nature we strive to overcome our nature?" Rubin asks. It means that what masquerades as love is actually our misery driving us to be something else. It is "a cattle-prod humanitarianism that has contempt for what we are in the name of the unfathomable things we could become."[19]

The capacity for transhumanist goals and ways of thinking to produce contempt for self and others is far from a minor point. On its website, Humanity+ claims that it is committed to "empathy as a way of life" but the ethical views that undergird its mission clearly work at cross purposes with that commitment. On the website one transhumanist insists that "Transhumanism is at heart altruistic. You are not working for your own account .... You work for all, for a future with more options, more freedom, more safety, less hunger, less poverty, less death. Ultimately it's about creating a world with more good and less bad in it."[20] But *The Transhumanist Wager* has made it clear that there is nothing within the transhumanist agenda—beyond a naive trust in enhancement drugs to make us more empathetic—that is consistent with persons *learning how* to feel empathy for others, or even to feel a basic connection with others except as conceived

[17] Ibid., 226.
[18] O'Connor, *Collected Works*, 831.
[19] Rubin, Charles. "Human Dignity and the Future of Man," 162.
[20] De Paus, "Transhumanists vs Fake Singularitarians."

abstractly. In a transhuman utopia, genuine connection with others is usually replaced with a crowd-sourcing, Spinozian-type of superintelligence. The only thing of value in that vision is your intellect in so far as it contributes to the whole.[21] What is not of inherent value is anyone's particular situation, anyone's particular pains, anyone's particular loves.[22] This is love for the future that is actually self-love, and a self-love that is actually contempt.

This trap of contempt is the primary reason why no transhumanist can write a compelling novel that emerges from this worldview. True storytelling is an art that carries within it a deep love for actual persons—persons who are, after all, worthy of a novel being written about them. Transhumanism neither understands how good character develops in persons nor recognizes that its vision for the good life is naive at best and downright ugly at worst. Insight into human experience, which is fiction's most powerful gift, can be no teacher to those who believe they already know what the future should look like. If we are incapable of learning why Tithonius got into trouble in the first place, we should not be surprised if we end up with his fate.

Bibliography

Atwood, Margaret. 2002. *Negotiating with the Dead: A Writer on Writing*. New York: Cambridge University Press.

Berry, Wendell. 2011. *Standing by Words: Essays*. Berkeley: Counterpoint.

Bostrom, Nick. 2008. Letter from Utopia. *Studies in Ethics, Law, and Technology* 2 (1): 1–7.

De Grey, Aubrey, and Michael Rae. 2007. *Ending Aging: The Rejuvenation Breakthroughs that Could Reverse Human Aging in Our Lifetime*. New York: St. Martin's Griffin.

De Paus, Koen. 2013. Transhumanists vs Fake Singularitarians. *H+ Media*, December 5. http://hplusmagazine.com/2013/12/05/transhumanists-vs-fake-singularitarians/

Gadamer, Hans-Georg. 2004. *Truth and Method*. New York: Continuum.

Harrison, Peter. 2015. *The Territories of Science and Religion*. Chicago: University of Chicago Press.

[21] The recent films *Lucy* and *Transcendence* reveal this fantasy very well.

[22] Peter Harrison (*Territories of Science and Religion*) explains how the Baconian revolution led to a new definition of Christian charity as the benefit and relief of man's state on earth.

Istvan, Zoltan. 2013. *The Transhumanist Wager*. Reno: Futurity Imagine Media LLC.

Meilaender, Gilbert. 2011. Thinking About Aging. *First Things: A Monthly Journal of Religion and Public Life* 212 (April 2011): 37–43.

Messerly, John G. 2014. Jaron Lanier on Transhumanism. *H+ Media,* July 7. http://hplusmagazine.com/2014/07/07/jaron-lanier-on-transhumanism/

Moylan, Tom. 1986. *Demand the Impossible: Science Fiction and the Utopian Imagination*. New York: Methuen.

Nussbaum, Martha Craven. 1990. *Love's Knowledge: Essays on Philosophy and Literature*. New York: Oxford University Press.

O'Connor, Flannery. 1988. *Collected Works*. Vol. 39. New York: Library of America.

Rubin, Charles. 2008. Human Dignity and the Future of Man. In *Human Dignity and Bioethics: Essays Commissioned by the President's Council on Bioethics,* ed. Edmund D. Pellegrino, Adam Schulman, and Thomas W. Merrill, 155–172. Washington, DC. https://repository.library.georgetown.edu/bitstream/handle/10822/559351/human_dignity_and_bioethics.pdf?sequence=1&isAllowed=y

Young, Simon. 2006. *Designer Evolution: A Transhumanist Manifesto*. Amherst: Prometheus Books.

CHAPTER 9

Do Bigger Brains Mean Smaller Gods? Cognitive Science and Theological Perspectives on Transhumanism and the Church (or, Why We Can't Outrun Faith)

Steve Donaldson

Hope, Hype, and Reality

Hope springs eternal in the human breast:
Man never is, but always to be blest:
The soul, uneasy, and confined from home,
Rests and expatiates on a life to come.[1] (Alexander Pope)

[1] Pope, *Essay on Man* (Epistle I), 7.

S. Donaldson (✉)
Samford University, Birmingham, AL, USA
e-mail: sfdonald@samford.edu

S. Donaldson, R. Cole-Turner (eds.), *Christian Perspectives on Transhumanism and the Church*, Palgrave Studies in the Future of Humanity and its Successors,
https://doi.org/10.1007/978-3-319-90323-1_9

Although it seems unlikely that many folks would deny Pope's sentiment, a bit of reflection suggests that, depending on the "hope" under consideration, one might also grant that some "hopes spring infernal." That an eternal hope could also be infernal has more than an alliterative ring to it, capturing as it does the idea that many hopes are in fact rooted in selfish desires. Frequently, however, the motivations may be mixed and difficult to decipher. Take, for example, the hope for eternal life itself—Pope's "life to come." It is hard to see any way in which this is not a selfish desire but also one that is encouraged in Christian traditions that otherwise decry selfishness. The fact that little is actually known about that life has posed no disincentive for Christians who, since the earliest days of Christianity, have even seen the mystery as part of the appeal. The assumption, of course, is that a God of love will make sure things work out—not in the end but in an eternity without one. However, for those unwilling to try the Christian experiment or who simply want to hedge their bets, other hopes for eternal life may appear attractive, especially if they seem to fall within the scope of what those individuals themselves (or the technologies they support) can muster. Similar age-old hopes for superior intelligence, enhanced physical prowess, or special skills present comparable eternal/infernal dilemmas for those who think that human efforts to obtain these things infringe upon God's turf.

Some Christians, however, are not so bothered, perhaps believing that enhancing one's cognitive and physical features are part of God's plan for us all along. Thus pitted against Friedrich Nietzsche's atheistic "übermensch" is the super Christian of Pierre Teilhard de Chardin's consummate noosphere.[2] That atheists and Christians could become such hopeful bedfellows regarding this future vision is perhaps even stranger than their joining ranks to oppose it, but the battle lines have been drawn. Whether we are stepping over our natural limitations or overstepping them, that is the question. Is it "No, no, no!" or "Go, go, go!"? For which side should we cheer?

Great Expectations: HOPE

Pope's refrain reflects the timeless desire among humans not only to live longer but to live better. Heaven would be nice, but why wait? And what does it mean to "live better"?[3] Although numerous responses are possible

[2] Teilhard de Chardin, *Phenomenon of Man*.
[3] Cf. Mitchell, "Define Better"; Bieber-Lake, *Prophets of the Posthuman*.

depending upon one's current condition—a vacation home at the beach would be nice, thank you, but so would clean drinking water if you currently have none—the transhumanist is thinking about gifts that are both more basic and more far-reaching. Thus the fruits of a transhumanist agenda would include cognitive enhancements which could enable clearer, more rational thinking, permit a greater understanding of the universe, and probe more deeply into life's big questions, thereby facilitating the acquisition of all those other pleasures that first come to mind when one thinks about "living better."

On the surface this sounds little different from any educational agenda. Yet this is not about education, but transformation, and if immortality is a product of these efforts, so much the better. In light of this, it is only natural to ask what types of cognitive enhancements might yield a better life. To answer this we could look at transhumanist aspirations but we could just as easily ask a student who is about to take an examination (perhaps for a course on the philosophical implications of transhumanist ventures!). It is not difficult to imagine the answers: total and instant recall would almost certainly be priorities, as would the ability to draw useful analogies,[4] avoid errors in logic, and express oneself coherently. These features which the student believes would contribute positively to good performance in the exam are of the same kind that should enable high achievement (i.e., better living) in all domains of life. (The individual who believes that wealth is the key to the good life, for example, would presumably welcome the cognitive means by which to acquire it.)

Reasons to think that the transhumanist hopes will be attained are quite visible. Aside from the obvious technologies employed by innumerable people every day, some advances are particularly telling. In an early book on transhumanism, Joel Garreau coined the now popular GRIN acronym for the technologies that would usher in the transhumanist age: Genetics, Robotics, Information, Nanotech.[5] Progress since then has been significant in each area. One has only to visit the publicly available website for the National Center for Biotechnology Information to appreciate the wealth of data and tools that have accumulated for analyzing genomes as a foundation for empowering a rapid increase in genetic understanding. Developments in prosthetics (which can be seen as adding robot-like features to humans) have been slower, but continue to progress, nonetheless.

[4] Cf. Hofstadter, *Fluid Concepts*; Kurzweil, *How to Create a Mind*; Bak, *How Nature Works*.
[5] Garreau, *Radical Evolution*.

When IBM's Watson defeated two of the world's best human Jeopardy players in 2011,[6] it was a significant step up from the triumph of the company's Deep Blue chess-playing program over human chess champion Gary Kasparov only a few years earlier[7]—a feat that involved substantial enhancements in information technology. Nanotechnology has already made possible the manipulation of individual atoms[8] as well as the creation of fanciful molecular structures in a process called DNA origami,[9] and projections for further progress by organizations such as the Foresight Institute are based on identifying "strategic research initiatives to deliver on this promise."[10]

All of this is quite important to remember because there is a tendency to lapse into science fiction when talking about transhumanism and to forget that not only is there real science behind the projections but also real progress. It is that kind of advance that has helped lead astrophysicist Neil deGrasse Tyson to declare that "[i]f I propose a God … who graces our valley of collective ignorance, the day will come when our sphere of knowledge will have grown so large that I will have no need of that hypothesis"[11]—a hope shared by many who are seeking any means possible to eliminate God from all of the gaps he is assumed to fill in human understanding. Of course among the various problems with this "God-of-the-gaps" view[12] are that it assumes (1) the only role of God is to serve as an answer to the questions we are asking (i.e., it ignores relational and salvific aspects of divinity); (2) we will find all the answers ourselves (and we will, of course, but will they be correct?); (3) we will ask the right questions (but we won't always, plus how will we know?); (4) God (if there is any at all) is really quite small.[13] In any case, there is more to the scientific story than is usually acknowledged, and by digging a bit deeper it is possible to see that some of the hope is mostly hype.

[6] Markoff, "Computer Wins on 'Jeopardy!'"

[7] Krauthammer and Dowell, "Kasparov: Deep Blue Funk."

[8] Eigler and Schweizer, "Positioning Single Atoms with a Scanning Tunnelling Microscope."

[9] Rothemund, "Folding DNA to Create Nanoscale Shapes and Patterns."

[10] Drexler et al., *Productive Nanosystems*, v.

[11] Tyson, Neil deGrasse, "Holy Wars: An Astrophysicist Ponders the God Question," in Kurtz, *Science and Religion*, 79.

[12] Cf. Drummond, *Ascent of Man*, 171.

[13] Reeves and Donaldson, *Little Book for New Scientists.*

Dreams of Sugar Plum Fairies: HYPE

The difference in hope and hype is often tenuous and there are important scientific reasons to think that some of the projected benefits of cognitive enhancement are considerably overstated. Here we'll focus on three examples.[14]

Computability

Carnegie Mellon roboticist Hans Moravec has argued that the primary factors holding back progress toward sentient machines are adequate processing speed and memory, limitations that he demonstrates are being reduced at an exponentially increasing rate by technological advancement. He has nicely illustrated this growth on a chart which also includes various animal life forms, whose raw brain capacities have been or are being overtaken by artificial systems. On that chart the processing power of human brains is depicted somewhere above mice and monkeys but below elephants and whales.[15] A moment's reflection is sufficient to note that the brains of the latter two animals are simply physically larger than human brains, thus dispelling any tendency to draw the erroneous conclusion that larger (i.e., faster, with greater memory capacity) means smarter. Yet that is the basic point Moravec wishes to make! What is missing from Moravec's chart (and he knows this) is that no amount of computing power alone is adequate for intelligence unless there is sufficient algorithmic content to accompany it.[16] The theme of exponential technological growth is played over and over by futurist Ray Kurzweil as the gateway not only to artificially intelligent systems but ultimately enhanced human cognitive abilities as well.[17] Kurzweil even attempts to provide an outline of a plan for machine intelligence,[18] following in a long succession of attempts to do so by a host of computer scientists who have approached

[14] Compare Grassie, "H-: Millennialism at the Singularity: Reflections on Metaphors, Meanings, and the Limits of Exponential Logic."

[15] Moravec, "When Will Computer Hardware Match the Human Brain?" See also Moravec, *Robot*.

[16] This is true both for any particular or general problem-solving strategies and for the problem-solving architecture itself.

[17] For example, Kurzweil, *How to Create a Mind* and *The Singularity Is Near*.

[18] Kurzweil, *How to Create a Mind*.

the subject from all sorts of angles.[19] At the time of this writing, Watson is arguably the most sophisticated product produced, although its powers are still quite limited.

The difficulty in producing sentient machines is not surprising, given the complexity of human brains that one wishes to emulate, but the issue about adequate algorithms may go beyond mere difficulty. The question that has been pondered for years is whether intelligence is actually algorithmic at all. Most computer scientists working in this area proceed under the assumption that it is, taking their cue from Alan Turing who discussed the matter in his seminal paper on computer intelligence in 1950.[20] Turing acknowledged that, mathematically, one could prove that there are problems which are not solvable by algorithm (the issue of "computability")—indeed he provided one of the primary conceptual tools, the Turing Machine, that is used today in undergraduate curricula to do so—but argued that it probably made no difference in human intelligence and would therefore be unlikely to do so in machine intelligence either. In fact, an algorithmic approach has even been proposed for emotions[21]—often thought to be a distinguishing feature of human intelligence and consciousness. The prevailing view is captured in Stephen Wolfram's *Principle of Computational Equivalence*: "All processes, whether they are produced by human effort or occur spontaneously in nature, can be viewed as computations."[22] But although both a provocative and useful insight, this merely begs the question—are intelligence and consciousness algorithmic?

For the British mathematical physicist Roger Penrose, the answer is "No" (at least for consciousness) and he uses one of Turing's concepts to argue why.[23] Like positions to the contrary, arguments that human artifacts cannot capture the essence of human cognition also have a long

[19] The list is enormous and the literature extensive. For a summary of some issues that must be addressed, see Donaldson, "Predictive Learning." Some noteworthy suggestions for tackling some of those problems include Albus, "Outline for a Theory of Intelligence"; Hawkins, *On Intelligence*; Donaldson, "A Neural Network for Creative Serial Order Cognitive Behavior."

[20] Turing, "Computing Machinery and Intelligence."

[21] Albus, "Outline for a Theory of Intelligence." Marvin Minsky (*Society of Mind*) acknowledged years ago that without emotions, machine intelligence would never rival that of humans.

[22] Wolfram, *A New Kind of Science*, 715–717.

[23] Penrose, *The Emperor's New Mind*.

history and adamant supporters.[24] Although the question seems more likely to be resolved in favor of machine intelligence than it does for machine consciousness, until one or the other is done the verdict is still pending. Part of the problem is that no one yet knows the extent to which consciousness might be involved in higher-order intelligence.[25]

Complexity

Besides computability, computer scientists also worry—or should—about something called computational complexity, a term used to denote the efficiency of an algorithm that does happen to exist. The relevance of this to transhumanist hopes is substantial, but seems to have been underplayed at best and ignored at worst. In a nutshell, the assumption has been that exponentially increasing computational power will enable solution of problems that are currently beyond the reach of existing technologies. That is surely correct, but it fails to note that the real problems of interest are exponentially more difficult than those which have been solved to date. Which curve has the faster rise? In other words, are computers with an exponential increase in computing power (which seems likely based on historic trends) going to be up to the exponentially more difficult tasks set before them?

A modest example will serve to illustrate the dilemma. Most people are familiar with a simple game called the 8-puzzle, which consists of a 3 × 3 grid of movable pieces numbered 1–8 with one blank space into which an adjacent piece can be moved either horizontally or vertically. The object is to begin with some scrambled board state (perhaps produced by a friend) and move pieces one at a time into the blank space until reaching some desired goal state (perhaps the pieces numbered consecutively from left to right, top to bottom, in which case the blank position would end up at the bottom right of the board). The puzzle can easily be conceptualized as a 4 × 4 grid instead (the "15-puzzle"), or even larger ("24-puzzle," "35-puzzle," etc.). Now it is fairly easy to write algorithms that can find an optimal solution to this problem (i.e., the shortest number of moves required to reach a goal state), but the efficiency of these algorithms can vary dramatically. For example, it is surprising to most folks to learn that it

[24] Cf. John Searle's famous "Chinese Room" thought experiment ("Minds, Brains, and Programs").

[25] That is, could a zombie be intelligent? Cf. Dennett, *Sweet Dreams.*

would take a relatively fast computer hundreds or even thousands of years to solve a moderately scrambled board for the 15-puzzle using a raw, brute force approach (which consists of examining all possible moves systematically until reaching the goal). Algorithmic approaches having this characteristic are termed intractable because they reflect something about a fundamental limitation of the approach and not merely the speed of the computer (i.e., significantly faster computers could still take eons to discover a solution). They may also indicate something inherently limiting about the nature of the problem itself and there are a host of problems that are currently deemed intractable in the sense that not only is there no known algorithm for efficiently solving them, there is good reason to believe that no such algorithms exist.

The 15-puzzle, however, is trivial, but for those significant but intractable problems of the type that transhumanists expect to be able to solve, optimal solutions will still remain out of reach. Whether it is computing the best possible move to make in chess,[26] trying to predict the protein (if any) that will result from a particular configuration of amino acids, or what physical or behavioral trait will result from a specific set of genetic instructions, gene interactions, and environmental constraints, faster computers still won't be the ultimate answer. In fact, the very argument used for claiming that computing power will reach a point beyond which we cannot predict what can be done can be applied to the problems themselves—until we have the enhanced power we cannot predict what the complexity of the new problems disclosed will be, but there is no reason to think that they will be less computationally challenging than their predecessors (most of which will also remain intractable). In short, we can no more foresee the problems than we can the promises,[27] and a "singularity"—the term popularized by Ray Kurzweil for that point in time beyond which we cannot project what will happen with exponentially expanding technologies[28]—is possible in either arena. The key point is that part of the

[26] Cf. Shannon, "Programming a Computer for Playing Chess."

[27] Some individuals think that quantum computing will be our savior in this regard. For example, Calude and Păun (*Computing with Cells and Atoms*, 253) report that Hughes has estimated that a factoring algorithm (used in encryption, perhaps) that would take years on a traditional Von Neumann computer could be done in minutes or hours on a quantum computer. It appears, however, that a quick solution to save cryptology is simply to increase the size of the keys used for encryption.

[28] Kurzweil, *The Singularity Is Near*.

problem with some of the hyped projections associated with transhumanist thinking is that they simply imagine enhanced brains dealing with the same problems we currently face.

Logic

In his excellent book on hemispheric differences in the brain, Iain McGilchrist gets a bit carried away with one of his analogies, noting that "it has been estimated that there are more connections in the human brain than there are particles in the known universe."[29] Now it is surely the case that there are a huge number of neural connections in human brains (McGilchrist's main point), but because every connection is mediated by particles (i.e., molecular components of ion channels and neurotransmitters), the number of connections in any brain must logically be a small subset of the total number of particles in that brain alone, much less the entire universe.

Work by scientists such as Thomas Gilovich, Amos Tversky, and Daniel Kahneman (to mention a few) illustrates various ways in which humans display logical reasoning errors,[30] but it probably doesn't take the observations of a scientist to convince most of us of this. One of Kurzweil's major hopes for a superintelligence is that it would not be prone to exhibit the kind of logical inconsistencies that now plague humans. In particular, he believes something on the order of a logical consistency checker could exist that eliminates such conundrums, many of which currently go undetected in the brains where they reside.[31] Now eliminating conflicting beliefs (what I have elsewhere called "polygamy of the thoughts"[32]) is a worthy goal, but although it is probably safe to say that some of those errors could be eliminated via cognitive enhancement, it is a mistake to think that all could and it may even be argued that a certain amount of logical tension is actually beneficial.

There are several reasons for claiming that it would be impossible to detect all logical inconsistencies. To begin, one of the reasons such inconsistencies arise in the first place is because the data on which attempted

[29] McGilchrist, *The Master and His Emissary*, 9.

[30] Gilovich, *How We Know What Isn't So* and "The Hot Hand in Basketball"; Tversky and Kahneman, "Judgment Under Uncertainty" and "The Framing of Decision and the Psychology of Choice."

[31] Kurzweil, *How to Create a Mind*, 176.

[32] Donaldson, *Dimensions of Faith*, 228.

logical conclusions are drawn are often incomplete. When that is the case, the only recourse is to make an educated guess to fill in the holes, but the guesses themselves are susceptible to the same problem of missing data in a recursive fashion that makes it impossible to guarantee consistency.

Furthermore, even if it is recognized that two (or more) beliefs are incompatible, it can be difficult if not impossible to know which is correct for the same reason given above. Consider, for example, a simple syllogistic chain of implications such as the following: A→B, B→C, C→D, D→E, and E→A′. The logical inconsistency of these statements is easy to see (i.e., both A and A′ cannot be true), but unless one can identify which of the implications are invalid, one can only guess about the logical relations between the statements. On which implication should the blame be placed? Making that decision may seem simple but, as just noted, it too can be subject to the same problem (i.e., each proposition shown above could be the result of its own syllogistic chain of reasoning or missing data). Note that this is not simply a matter either of the number of supporting hypotheses or of the most direct logic trail.

The example above involved deductive reasoning but the premises used are frequently the result of an inductive process, which by its very nature is not logically sound. Even if it is possible to reduce the number of inductive errors, there will never be a way to eliminate them because it is often only after the passage of sufficient time and experience that some of those inductive inferences can be judged faulty. On the other hand, there can never be enough time to judge an inductive inference universally true (unless all cases can be exhausted) and there is consequently always the potential for a new observation to refute a cherished belief.[33] Many search spaces are quite large and can only be sampled so that all conclusions are really just probabilistic anyway. Even the samples themselves might change from moment to moment (e.g., if we are talking about the beliefs of others, for instance), and apparently good reasons for believing something can therefore fluctuate with time as well. Finally, it is a well-established fact in mathematics that even in a system that is logically sound, there will always be propositions that cannot be proven within that system[34]—the computability issue once again.

It is also important to note that a logic consistency checker may be unable to escape the intractability issues previously discussed. The more

[33] Cf. Taleb, *Black Swan*.

[34] Gödel, "Über Formal Unentscheidbare."

powerful an intelligence becomes, the more constraints it will presumably employ when attempting to verify consistency, but that means potentially bigger search trees with correspondingly longer search times. Perhaps new heuristics will be discovered to speed the search but relying strictly on that hope is to neglect the fact that new problems will always exist for which heuristics have not yet been found or simply do not exist. And so it goes...

So far we have explored several reasons for thinking that the detection of all logical inconsistencies is a pipe dream but one can also wonder if there might be situations where failing to recognize logical inconsistencies or an inability to correct them might be essentially beneficial or have at least some redeeming quality. To address this we need to distinguish between those inconsistent beliefs of which we are conscious and those of which we are not. Now on the surface it would seem desirable to become conscious of all unconscious yet inconsistent beliefs, if for no other reason than that some of the inconsistencies could manifest themselves in non-productive, perhaps foolish, and possibly even dangerous attitudes and behaviors. Raising those beliefs to a level of conscious awareness offers the potential for us to make rational attempts to correct them. However, because they are unconscious, we have no way of knowing how many such beliefs we entertain. If the number is few, then bringing them to light and correcting them seems like it should be a priority. But what if there are many? We know much about the world, having acquired that knowledge from a variety of sources which have themselves been influenced by other sources, all within their respective socio-cultural climates, and the possibility that many of the acquired assumptions, ideas, and inferences clash at some fundamental level seems quite real. If they were all to suddenly enter our consciousness, how should we decide which deserves our limited attention? Even that decision itself might rest on a collection of assumptions and conclusions which have slipped into our subconscious minds. Consequently, it might be that even if holding inconsistent beliefs is not desirable in itself, remaining unaware of all but the most important could actually contribute to our sanity. For a superintelligence which (as discussed above) cannot escape having inconsistent beliefs any more than we do but which experiences each begging for resolution, the results could be paralyzing. As Wendell Berry has asked, "[i]f you had complete knowledge, if you knew everything, could you then act? Could you apply what you knew, or would you be paralyzed by a surplus of considerations?"[35]

[35] Cf. Berry, *Life Is a Miracle*, 149.

One solution, of course, is simply to assume a prioritization mechanism. Prioritization by itself, however, cannot eliminate conflicts and it actually adds to the number of factors that must be considered. Although an omniscient and omnipotent God might escape this dilemma, doing so for finite creatures of whatever capacities seems unlikely.

None of this is to deny the desirability of seeking to discover and eliminate inconsistent beliefs, and for individuals who recognize the threat of holding such beliefs there will usually be a conscious attempt to identify and remove them. Yet the impossibility of eliminating all of those need not be the nightmare that might be imagined. Because there will always be aspects of the world for which different explanatory programs are possible,[36] short of omniscience, those explanatory frameworks will sometimes be in conflict. As the Princeton theologian J. Wentzel Van Huyssteen (citing Nicholas Rescher) reports, competing rationalities and their associated behaviors can exist with different justifications, albeit good reasons, for each.[37] The very presence of those competing rationalities can be the motivation which not only drives us to search for reconciling solutions but also to try to see the larger picture into which they fit.

In short, knowing that both we and our transhuman successors are susceptible to logical inconsistencies can motivate each of us to search for and resolve them. However, believing that we can eradicate them entirely (or have already done so) is more hype than hope and will simply generate complacency and a lack of empathy for our fellow strugglers, human or transhuman.[38]

We Can't Outrun Faith: REALITY

Undoubtedly, some of the hype will become a reality—but we still won't be God. Much in yesterday's science fiction has become today's science and there is no reason to think that trend will not continue. However, attempting to acquire the image of any deity with infinite attributes and knowledge will be like trying to put an ocean into a bigger cup when it is discovered that it will not fit into a smaller one.

[36] Cf. McGrath, *Science and Religion*, 51–58; Rosenberg, *Philosophy of Science*, e.g., 39–59.

[37] Van Huyssteen, *Shaping of Rationality*, 157.

[38] See McGilchrist (*The Master and His Emissary*) regarding concerns about an agenda driven strictly by one logical perspective (left brain) to the exclusion of a larger logical, empathic framework (right brain).

Problems such as those outlined above which demonstrate the incompleteness of cognitive capabilities even in a transhuman future are another way of saying that we can't outrun (or outgrow) faith. This will surprise (and possibly be denied by) many individuals who believe (an act of faith!) that greater intelligence will move us and our transhuman descendants from the realm of faith to fact. This view, however, misapprehends both the nature and role of faith and its inescapable operation in the brains of any creatures short of deity.[39] The fact is, however, that faith is necessitated by a variety of human limitations,[40] the interpretation of approximate solutions to intractable problems,[41] plus the compounding factors of a universe where random events are and will be the norm[42] and where there is a sensitivity of dynamic systems to initial conditions.[43] "Better" brains might help but they provide no ultimate escape.

When faith is understood in a normative way as "the probability assigned by an individual to the existence of something (including the nature of its attributes) or to the occurrence of an event (past, present, or future), or to the reason for the existence or occurrence of something,"[44] the inescapable nature of faith in us and our descendants becomes easy to see. Furthermore, it becomes apparent that faith is not limited to any particular domain—it is operative in every arena of existence inhabited by finite brains.[45] Consequently, any supposed knowledge held by a superintelligence will still have a probabilistic character, meaning such a being will still have beliefs, and decisions made on the basis of those beliefs will always be subject to error.

Christians (and others!) frequently balk at this approach, but it not only captures the essential nature of faith, it also helps remove the nebulous and mysterious elements often wrapped up in traditional notions, especially about "religious" faith. Faith of any kind is always about (or directed toward) something (the object of the faith) which may or may not be

[39] And, as an Open Theist might argue, possibly even there as well.

[40] Donaldson, *Dimensions of Faith*, 79–112.

[41] Approximation techniques can dramatically speed the discovery of solutions for intractable problems but without being able to guarantee optimality. Such losses of certainty are integrally involved with the concept of faith.

[42] Whether that randomness is ontological (real) or epistemological (perceived due to limited capabilities) makes little difference when it comes to human or transhuman faith.

[43] Kauffman, *At Home in the Universe*, 16–17, 23.

[44] Donaldson, *Dimensions of Faith*, 47.

[45] Ibid., 17, 29–31, 33, 47, 65.

important (the significance of the object). The probability associated with that faith is just a measure of its strength (seldom considered explicitly) and the accuracy of the probability reflects the merits of the faith.[46] Attitudes and behaviors based on faith (e.g., from marriage to martyrdom) involve all of these components and thus reflect something about the cognitive content in the minds (and hence brains) of individuals. For humans and their transhuman descendants, this can be summed up concisely as "no faith, no brain." Note that this is dramatically different from a fairly common derogatory view that "faith equals no brain."

So, while a desire for perfection can be deemed a Christian responsibility and a target for which to shoot,[47] its achievement will always remain out of reach. Therefore, clarifying the goals of perfection and correcting the motives for striving to attain it—that is, is one trying to be God or just to have more of his image?—will remain viable ministries of the church even in a transhuman future. But in addition to meeting spiritual needs, the church has also always been concerned with addressing physical, emotional, and educational needs. Yet even when motives are pure and intentions good, some of those needs may currently be beyond reach. Although it does not require a transhumanist perspective to attempt to better the human condition in these areas—indeed empathic people have been trying to do that all along—in this case transhumanist goals for better brains could be the gateway to make some of those good intentions possible. Recognizing such possibilities, drawing attention to them, and helping implement them will remain a major calling for the church of tomorrow.

Ultimately, however, cognitive enhancements may make no significant differences in whether one does or does not become a Christian or whether one attempts to become more Christ-like.[48] It is fashionable to believe that greater education weakens religious convictions, but even if that is the case, it is probably true only up to a certain point—extraordinarily bright and well-educated people through history since the time of Christ have been Christians. For the upgraded brains of transhumanist hopes there may be an ability to see more deeply into issues and consider more information but, ultimately, the need for faith will manifest itself in a variety of choices essentially like those observed today.

[46] Ibid., 114, 223–36.

[47] Matt 5:48.

[48] In possible contrast to Teilhard's vision (*The Phenomenon of Man*).

Christians might reasonably imagine that an ability to see deeper into the gospel message (enabled by enhanced cognitive abilities) would make the decision to accept it easier for those on the fence, but that same ability could persuade others to reject or abandon it. Enhanced brains will continue to be able to find evidence for one point of view or the other but, as with current brains, much of that supposed search for evidence will frequently be to support an existing bias. As Iain McGilchrist puts it, "[t]he nature of the attention one brings to bear on anything alters what one finds."[49] Thus, superior brains will still have their prejudices. Furthermore, even in the face of evidence, there will be no escape from the same factors that currently act as deterrents to acceptance (e.g., pride, unwillingness to acknowledge sinfulness and to repent, desire for immediate gratification).[50] Better sight does not necessarily mean better insight.[51]

This suggests that there may be differences in reasons for not wanting to hold incompatible beliefs or in what is thought to be the real value in not doing so: for some the value is to be able to come to a knowledge of the truth, which may be deep and multifaceted and require merging beliefs from different arenas of knowledge and experience; for others it may be more about finding an assumed justification for rationalizing something already believed.[52] Actually, both goals can be found in the same brain today at different times or for different issues. Ideally our super-progeny would lean toward greater insight but of that there is no guarantee if those beings remain free to choose.

The extent to which such creatures are really free remains unknown. Perhaps it will be no clearer to them than it is to us; maybe they will see more deeply. C.S. Lewis has suggested that engineered superhumans will have lost their freedom[53] but that is not a necessary conclusion. After all, whether God, evolution, or a happy mixture of the two is responsible for engineering us, fundamental questions about freedom are no less real now than they will be in the future. In the sense of enabling possibilities, it

[49] McGilchrist, *The Master and His Emissary*, 29.

[50] Cf. Peters, "Progress and Provolution: Will Transhumanism Leave Sin Behind?" in Cole-Turner, *Transhumanism and Transcendence*.

[51] An attempt to reclaim the perfections of Adam is sometimes cited as one of the driving factors behind the development of modern science (Harrison, *The Bible, Protestantism, and Rise of Natural Science*), yet one of the primary conclusions regarding the Fall is that perfect sight does not equal perfect insight.

[52] Cf. the explorer/mechanic metaphor in Donaldson, *Dimensions of Faith*, 139–167.

[53] Lewis, *The Abolition of Man*.

seems safe to say that technological enhancements will free our offspring to do things we cannot. But in the more noteworthy sense of seeing freedom as a characteristic of genuine autonomy, the vision is less clear. All creatures exist within some environmental context that shapes who and what they become, and such contexts for future generations will be just as operative then as they are today.

The relationship between freedom, the features of creatures occupying a given environment, and the environment itself is therefore a tenuous one. Both we and our super-offspring, therefore, live (and will live) on a perpetual border between freedom and determinism. Nevertheless, it ultimately appears that biblical injunctions to be loving, forgiving, and otherwise Christ-like only make sense in light of some degree of personal autonomy.[54] That is also true of calls by individuals (a number of whom otherwise claim not to believe in free will[55]) for others to think or behave in some way they deem desirable—those calls themselves are only comprehensible if one is free to heed them.

Having freedom that our descendants lack would, as Lewis fears, indicate that humanity has not stepped forward but backward.[56] Ideally, any viable transhuman offspring should possess at least as much freedom as its predecessors. Yet any true freedom to choose means the freedom to choose poorly, and this in turn suggests that the church will still be about business as usual in helping to guide the formation of moral and ethical decisions—a matter particularly germane to this discussion because the church's message that "the truth will set you free"[57] is inseparable from the idea that poor choices are intimately connected to an ultimate loss of freedom.

Furthermore, if being made in God's image implies anything (and it probably implies much),[58] then the creativity required for transhumanist efforts might be seen as reflecting a portion of that image.[59] Whether such

[54] Although biblical positions on free will may seem ambiguous, this is one possible avenue of resolution.

[55] Cf. Wegner, *The Illusion of Conscious Will*; Wilson, *Consilience*.

[56] Lewis, *The Abolition of Man*.

[57] John 8:32.

[58] Cf. Grenz, *The Social God and the Relational Self*; Herzfeld, *In Our Image*; Van Huyssteen, *Alone in the World?*; Shults, *Reforming Theological Anthropology*; Jensen, *Systematic Theology*.

[59] This is because claims for the *imago Dei* follow immediately after the portrayal of God's own creative acts.

creative acts actually do so could depend upon a vision of the church in which it foresees part of its role as helping to shape the motives and methods employed (as opposed to simply labeling the whole effort "misguided").[60] If creativity is indeed part of the *imago Dei*, then it may also be that choosing to be creative is one of the highest expressions of our (and our descendants') freedom. Conversely, it may be that very creativity which provides the strongest evidence for our freedom[61] because, without creativity, attitudes and actions tend to be deterministic and stereotyped.

This is not to deny that there are creative acts that could lead to the loss of freedom for others.[62] But it is just as plausible that the reverse is true as when, for example, one discovers a cure for the debilitating lack of freedom such as that seen in such maladies as autism, Lesch–Nyhan syndrome, hemi-spatial neglect, and so on. Thus it is interesting to contemplate the possibility that there may be a synergistic relationship between creativity and the enhancement of freedom.

Conclusion

The prospects for achieving all transhumanist goals are as yet still vague but some level of attainment is highly likely. For instance, engineered immortality may be a long way off (perhaps an eternity?) but significant life extension seems quite plausible. Similarly, fully integrated brain–machine systems that extend human capabilities will probably develop in bits and spurts but slowly become evermore powerful. Purely physical enhancements are probably much more realistic. As previously mentioned, concerns about lapsing into science fiction are genuine because they tend to demean the real science that is occurring and the actual prospects for some portion of the transhumanist agenda to come to fruition. Nevertheless, there are good reasons for claiming that some of the projected improvements are more hype than hope.

With that in mind, it is useful to note that regardless of the extent of transhuman development over the next several years or decades, thinking about these matters can help provide new and productive frameworks for

[60] This positive approach is likely what Teilhard (*The Phenomenon of Man*) had in mind in his vision.

[61] Cf. Donaldson, *Dimensions of Faith*, 220, 244–245, where exploration and discovery (both of which can be associated with creativity) are portrayed as integral to freedom.

[62] Cf. the concern of C.S. Lewis in *The Abolition of Man*.

approaching the deep issues with which the church has always been concerned. At the same time, those frameworks will illuminate new concerns which the church could fail to heed at its peril.

The conundrum of how to prepare for an unknown future can be addressed in two ways. One is to attempt to imagine all of the possibilities for such a future and to suggest potential responses to each. That is a potentially worthwhile approach for the short term but probably not very productive for the longer term. The second option (which can supplement the first) is to help shape that unknown future. Churches will do this by not denying the science or the potential usefulness of transformation—they will happen anyway—and by trying themselves to be logically consistent in their approaches to these issues. For example, many Christians opposed to the transhumanist agenda on the basis that radical intervention is unacceptable in the physical realm believe radical intervention in the spiritual realm is not only acceptable but required. If transhumanism can help us acquire more of God's image, then we should at least acknowledge that as a potential benefit.

The Christian is instructed to "be transformed by the renewal of your mind."[63] Today that occurs in ways the apostle Paul never envisioned via a variety of techniques ranging from the use of drugs to neurosurgery to brain–machine interfaces. Nevertheless, the transformations that occur with those approaches are all involved in who and what we are and our descendants will be. To remain viable, the church will need to minister to those descendants just as well as it ministers to its members and prospects today. As we have seen, those future entities will still be creatures of faith, and as such will be in need of Christian attention no less than current individuals. That those ministries will require creativity seems assured.

Bibliography

Albus, James S. 1991. Outline for a Theory of Intelligence. *IEEE Transactions on Systems, Man, and Cybernetics* 21 (3): 473–509.

Bak, Per. 1999. *How Nature Works: The Science of Self-Organized Criticality.* New York: Springer.

Berry, Wendell. 2001. *Life Is a Miracle: An Essay Against Modern Superstition.* Berkeley: Counterpoint.

[63] Romans 12:2.

Bieber-Lake, Christina. 2013. *Prophets of the Posthuman: American Fiction, Biotechnology, and the Ethics of Personhood.* Notre Dame: University of Notre Dame Press.

Calude, Cristian S., and Gheorghe Păun. 2001. *Computing with Cells and Atoms: An Introduction to Quantum, Membrane, and DNA Computing.* New York: Taylor and Francis.

Cole-Turner, Ronald, ed. 2011. *Transhumanism and Transcendence: Christian Hope in an Age of Technological Enhancement.* Washington, DC: Georgetown University Press.

Dennett, Daniel C. 2005. *Sweet Dreams: Philosophical Obstacles to a Science of Consciousness.* Cambridge, MA: MIT Press.

Donaldson, Steve. 1999. Predictive Learning and Cognitive Momentum: A Foundation for Intelligent, Autonomous Systems. Proceedings of the 37th Southeast Regional Conference of the ACM.

———. 2008. A Neural Network for Creative Serial Order Cognitive Behavior. *Minds and Machines* 18 (1): 53–91.

———. 2015. *Dimensions of Faith: Understanding Faith Through the Lens of Science and Religion.* Eugene: Cascade.

Drexler, Eric, John Randall, Stephanie Corchnoy, Alex Kawczak, and Michael L. Steve, eds. 2007. *Productive Nanosystems: A Technology Roadmap.* Palo Alto: Foresight Institute.

Eigler, D.M., and E.K. Schweizer. 1990. Positioning Single Atoms with a Scanning Tunnelling Microscope. *Nature* 344: 524–526.

Garreau, Joel. 2006. *Radical Evolution.* New York: Broadway.

Gilovich, Thomas. 1991. *How We Know What Isn't So: The Fallibility of Human Reason in Everyday Life.* New York: Free Press.

Gilovich, Thomas, Robert Vallone, and Amos Tversky. 1985. The Hot Hand in Basketball: On the Misperception of Random Sequences. *Cognitive Psychology* 17: 295–314.

Gödel, Kurt. 1931. Über Formal Unentscheidbare Sätze der Principia Mathematica und Verwandter Systeme, I. *Monatshefte für Mathematik und Physik* 38: 173–198.

Grassie, William. 2011. H-: Millennialism at the Singularity: Reflections on Metaphors, Meanings, and the Limits of Exponential Logic. *Metanexus,* August 9. http://www.metanexus.net/essay/h-millennialism-singularity-reflections-metaphors-meanings-and-limits-exponential-logic

Grenz, Stanley. 2001. *The Social God and the Relational Self: A Trinitarian Theology of the Imago Dei.* Louisville: Westminster John Knox Press.

Harrison, Peter. 2001. *The Bible, Protestantism, and the Rise of Natural Science.* Cambridge: Cambridge University Press.

Hawkins, Jeff, and Sandra Blakeslee. 2004. *On Intelligence*. New York: St. Martin's Griffin.

Herzfeld, Noreen L. 2002. *In Our Image: Artificial Intelligence and the Human Spirit*. Minneapolis: Fortress.

Hofstadter, Douglas, ed. 1995. *Fluid Concepts and Creative Analogies*. New York: Basic.

Jensen, Robert. 2001. *Systematic Theology: The Works of God*. Vol. 2. New York: Oxford University Press.

Kauffman, Stuart. 1996. *At Home in the Universe: The Search for the Laws of Self-Organization and Complexity*. New York: Oxford University Press.

Krauthammer, Charles, and W. Dowell. 1996. Kasparov: Deep Blue Funk. *Time*, February 26.

Kurtz, Paul. 2003. *Science and Religion: Are they Compatible?* New York: Prometheus.

Kurzweil, Ray. 2006. *The Singularity Is Near: When Humans Transcend Biology*. New York: Penguin.

———. 2013. *How to Create a Mind: The Secret of Human Thought Revealed*. New York: Penguin.

Lewis, C.S. 2001. *The Abolition of Man*. New York: HarperOne.

Markoff, John. 2011. Computer Wins on 'Jeopardy!': Trivial, It's Not. *The New York Times*, February 16. http://www.nytimes.com/2011/02/17/science/17jeopardy-watson.html?pagewanted=all&_r=0

McGilchrist, Iain. 2012. *The Master and His Emissary: The Divided Brain and the Making of the Western World*. New Haven: Yale University Press.

McGrath, Alister. 2010. *Science and Religion: A New Introduction*. Malden: Wiley-Blackwell.

Minsky, Marvin. 1985. *The Society of Mind*. New York: Simon and Schuster.

Mitchell, Ben. 2004. Define Better. *Christianity Today* 48 (1): 42–44.

Moravec, Hans. 1998. When Will Computer Hardware Match the Human Brain? *Journal of Evolution and Technology* 1. http://www.jetpress.org/volume1/moravec.pdf

———. 1999. *Robot: Mere Machine to Transcendent Mind*. New York: Oxford University Press.

Penrose, Roger. 1989. *The Emperor's New Mind: Concerning Computers, Minds and the Laws of Physics*. Oxford: Oxford University Press.

Pope, Alexander. 1844. *An Essay on Man in Four Epistles*. Troy: W. & H. Merriam.

Reeves, Josh, and Steve Donaldson. 2016. *A Little Book for New Scientists: Why and How to Study Science*. Downers Grove: IVP Academic.

Rosenberg, Alex. 2012. *Philosophy of Science: A Contemporary Introduction*. New York: Routledge.

Rothemund, Paul. 2006. Folding DNA to Create Nanoscale Shapes and Patterns. *Nature* 440: 297–302.

Searle, John R. 1980. Minds, Brains, and Programs. *Behavioral and Brain Sciences* 3 (3): 417–457.

Shannon, Claude E. 1950. Programming a Computer for Playing Chess. *Philosophical Magazine* 41 (314): 256–275.

Shults, F. LeRon. 2003. *Reforming Theological Anthropology: After the Philosophical Turn to Relationality*. Grand Rapids: Eerdmans.

Taleb, Nassim Nicholas. 2007. *The Black Swan: The Impact of the Highly Improbable*. New York: Random House.

Teilhard de Chardin, Pierre. 2008. *The Phenomenon of Man*. New York: Harper Perennial.

Turing, Alan. 1950. Computing Machinery and Intelligence. *Mind* 59: 433–460.

Tversky, Amos, and Daniel Kahneman. 1974. Judgment Under Uncertainty: Heuristics and Biases. *Science* 185 (4157): 1124–1131.

———. 1981. The Framing of Decision and the Psychology of Choice. *Science* 211 (4481): 453–458.

Van Huyssteen, J. Wentzel. 1999. *The Shaping of Rationality: Toward Interdisciplinarity in Theology and Science*. Grand Rapids: William B. Eerdmans.

———. 2006. *Alone in the World?: Human Uniqueness in Science and Theology*. Grand Rapids: Eerdmans.

Wegner, Daniel. 2003. *The Illusion of Conscious Will*. Cambridge, MA: Bradford.

Wilson, Edward O. 1999. *Consilience: The Unity of Knowledge*. New York: Vintage.

Wolfram, Stephen. 2002. *A New Kind of Science*. Champaign: Wolfram Media.

PART III

Technology and the Church of the Future

CHAPTER 10

Will the Transhuman Future Be Good or Bad for Humanity?

George Michael Glawson

Introduction

Ray Kurzweil is one of the most brilliantly successful inventors and businessmen living today. He has received 20 honorary doctorates and the National Medal of Technology, was principal inventor of many of the technologies we use every day, and is currently (among other things) the Director of Engineering at Google. In a recent interview, when one might have expected him to focus on the accomplishments of his life, he talked instead about an issue of personal importance, the matter of death:

> I have a recurring dream that has to do with exploring this endless succession of rooms, going from one to the next, and feeling hopelessly abandoned and lonely—unable to find anyone else. That's a pretty good description of death. Death is supposed to be a finality, but it's actually a loss of everyone you care about. I have fantasies sometimes about dying, about what people

G. M. Glawson (✉)
University of South Carolina, Columbia, SC, USA

S. Donaldson, R. Cole-Turner (eds.), *Christian Perspectives on Transhumanism and the Church*, Palgrave Studies in the Future of Humanity and its Successors,
https://doi.org/10.1007/978-3-319-90323-1_10

> must feel like when they're dying, or what I would feel like if I were dying, and it's such a profoundly sad, lonely feeling that I really can't bear it. ... So I go back to thinking about how I'm not going to die.[1]

Kurzweil, in addition to his many other roles, is also one of the intellectual leaders of the transhumanist movement—a movement whose various aims often sound like the stuff of science fiction: the merging of bodies and machines, the interfacing of brains and computers, the enhancement of senses or the creation of entirely new ones; the list could go on. These proposals beg for deep ethical reflection. In this chapter, I focus on two general questions: First, would a future that the transhumanist movement strives for be good or bad for humanity? Second, what is the Church's role in periods of radical technological change?

The Nature of Human Nature

In order to ask clear, probing questions about transhumanism, we need a clearer concept of it. Nick Bostrom, one of the intellectual leaders of the movement defines transhumanism in this way:

> Transhumanists view human nature as a work-in-progress, a half-baked beginning that we can learn to remold in desirable ways. Current humanity need not be the endpoint of evolution. Transhumanists hope that by responsible use of science, technology, and other rational means we shall eventually manage to become posthuman, beings with vastly greater capacities than present human beings have.[2]

An Intuitive Argument Against Transhumanism

The natural reaction to the idea of transcending our nature using technology is a suspicion and severe ambivalence—feelings that we do not usually have toward technological projects with which we are more familiar. Our suspicion and fear is often centered on the idea that these technological changes aim to do something radically new to us. In contrast to familiar everyday technologies which simply act as tools for our own chosen ends, transhuman technologies aim to fundamentally alter our most basic human qualities—our bodies, our senses, how we engage in the most

[1] Kurzweil, Ray, *Transcendent Man*.
[2] Bostrom, Nick. "Transhumanist Values."

basic human activities. This seems to cross a line that past technologies have kept to one side of: the line between technologies that exist externally, *for* us, and technologies that have *invaded* our being and thus cease to be merely external use-objects. Transhumanism does not simply aim to change the world of technology and tools that we inhabit; it seeks to alter *humanity itself*.

This argument against transhumanism has a strong intuitive appeal, but it contains two important, and false, assumptions—one about the nature of humans and another about the nature of technology. The first false assumption is that, by mingling with human nature at the biological level, transhuman technologies would be doing something radically different from other, common technologies. The second false assumption is that our everyday technologies are essentially passive tools that exert no profound influence on humanity since they are not deeply, biologically integrated with us the way many transhuman technologies would be.

The falsity of these assumptions can be seen best indirectly, by taking a closer look at the relationship between human nature and technology. To this we will now turn.

Human Nature: Homo ________

The standard term for the human species, *homo sapiens*, was not the obvious or uncontested choice for our species' scientific name. This is because the scientific name often aims to capture the ultimate distinguishing feature of the kind and, when it comes to humans, picking one characteristic feature is hard to do. In the end, *homo sapiens* prevailed. The term *sapiens* means "knowing" or "wise," and selects as our distinguishing feature the human abilities to retain knowledge and reason about what we know. Nevertheless, other competing terms that point to different defining human traits still get occasional use, each pointing to some important aspect of human nature (i.e., some quality without which we would not be characteristically human).[3]

[3] These include "homo faber" (Arendt, *The Human Condition*); "homo ludens" (Huizinga, *Homo Ludens*); "homo sentimentalis" (Halton, *Bereft of Reason*); "homo socius" (Berger and Luckmann, *The Social Construction of Reality*); "homo domesticus" (Jensen, *Endgame*); "homo animalis" (Martin Heidegger, "Letter on Humanism," in Heidegger, *Basic Writings*); "homo religiosus" (Alister Hardy, as quoted by Rüdiger Vaas, "God, Gains, and Genes," in Voland and Wulf Schiefenhövel, *The Biological Evolution of Religious Mind and Behavior*); "homo poetica" (Becker, *The Structure of Evil*).

Given these many different dimensions of our nature, perhaps the most accurate description is ***homo duplex*** ("divided man"), a term proposed by Émile Durkheim to suggest that what distinguishes humanity is not any *one* quality, but the fact that our natures are irreducibly multidimensional. This makes intuitive sense—we would view a person born without the capacity for language, emotion, religion, relationships, or creativity as missing something just as fundamental to being human as a person born without a normal human biology (*biologicus*) or human intelligence (*sapiens*).

This point about the irreducible multidimensionality of human nature hints at the first faulty assumption of the argument against transhumanism laid out above. That argument relies on the idea that, by integrating with our bodies and altering our senses, transhuman technologies are tampering with human nature to a degree that "normal" technologies do not. However, this relies on the idea that our biological qualities are somehow *more sacred*, more fundamentally a part of our nature, than our other qualities. However, we are not fundamentally just *homo biologicus*, or *homo poetica*, or any other; we must view all these dimensions as equally fundamental. The argument against transhumanism, therefore, holds a shallow biological conception of human nature and ignores the other equally essential dimensions of humanness.

With respect to the mingling of technology with human nature, the genie is already out of the bottle. Technologies have integrated with and radically transformed *homo sapiens*—we learn, remember, and think differently now that we use technologies like writing, computers, digital data storage, the Internet, and calculating machines. We have also technologically transformed *homo sentimentalis*. Our emotions are now aroused by and developed in response to our experience of events and characters, both real and fictional, as they come to us through the screens, radios, and novels that alter them and our emotional response to them in important ways.

One could keep going down the list, enumerating the ways various dimensions of human nature have been radically transformed over time through technology. We have seen enough to make the point though: human nature has already been altered technologically, and yet we do not seem to have wound up in a dystopic transhuman future. The particular alterations to human nature proposed by transhumanism may be novel, but technological alterations to fundamental aspects of our nature are not, and while technologies of the near future seem radically different from earlier human tools, the difference is largely a matter of degree. Laser-assisted surgery appears to be miles apart from scraping a wound with a rock, but it is hard to show a qualitative difference in *how* we relate to those technologies. Our tools have gotten more precise, more complicated, and are built in accord with more rigorously developed theories about the world, but these are only matters of degree. We have always been using the tools we have to alter ourselves. In a very real sense, we have always been transhuman.

Technological Nature

We can now turn to the second problematic assumption of the negative argument. This is the idea that proposed transhuman technologies exert a unique influence over us because they become active *parts* of us, unlike other technologies, which are merely *passive* tools that we use when we need, then put aside. To see whether this assumption is true, we need to consider our historical relationship to our technologies.

We have used tools (a term I will use interchangeably with *technologies*) from our very beginning. In fact, one of the popular accounts of human nature in anthropology textbooks conceives of humans as *homo technologicus*, that is, as essentially tool users. Even if our species is not unique in its use of tools, a penchant for creating and using them is certainly one more prominent dimension of our nature. This is why, when we read (always fictional, or mistaken) accounts of biologically human groups who live without any tools at all (*homo biologicus* who are not *homo technologicus*), we inevitably think there is something *sub*human about them. They are missing something essential to what it means to be one of us, for humans *by nature* are tool users.

If this seems far-fetched, simply consider God's first command that humans should "fill the earth and subdue it; and have dominion over ... every living thing that moves upon the earth."[4] If this is taken as an indication of our natural purpose, and therefore of our nature, then we are by nature always caught up with the use of the tools and technologies necessary for the task of ruling over and cultivating the land—something impossible to do with one's bare hands. Consider further that the basic activities characteristic of human life such as hunting, agriculture, construction of dwellings, trade, and communication all require the use of certain

[4] Genesis 1:28 (New Revised Standard Version).

technologies. Tools are therefore not mere after-the-fact *accessories* to the natural human life, but *necessities* for it. The earliest humans lived thoroughly technological lives, just as we do, and it makes no difference that their technologies seem as primitive to us as ours will to future humans.

By recognizing the fact that humans are essentially technological—essentially tool users—we open ourselves to an understanding of technologies as our longstanding partners, rather than after-the-fact accessories or invaders from without. This is not to suggest that we should not be critical of our technological companions. Rather, it simply helps us avoid engaging in wrongheaded criticisms. We know that it is not our *closeness* to technologies that is problematic, since we have always depended on technologies to pursue the fulfillment of our natures. It is rather the *particularities* of how our tools affect us that we should scrutinize. To this end, we should take a moment to see just how this partnership between humans and their technologies works, especially with regard to the ways our technologies exert their own influence on us. This will help us see more problems with the assumptions that non-transhuman technologies are more innocuous or passive than transhuman technologies, and that they exert less of an influence over who and what we are.

Technology and Morality

Technological Influence on Our Values and Aims

We begin with a familiar claim: *technologies in themselves are neither good, nor bad, but capable of being put to ends both good and bad.* The pen can be used to express either cruelty or kindness. The blade can kill or defend. And so the character of the end to which any technology, tool, or artifact is put is determined largely by the character of the person who wields it. Does this not suggest that everyday technologies are neutral tools over which we have total control, unlike transhuman technologies which integrate with, and exert influence over, us?

This picture seems commonsensical, but ignores subtle and profound ways *all* tools exert influence over their users and the societies in which they are used. Tools are not merely passive instruments that we put to use for our independently formulated aims, then set aside. They are in fact active players in the complex system of our values, aims, plans, and actions, they encourage us toward particular activities and ends, and they foster subtle biases. This is the point of the saying that *everything looks like a nail to the one who wields the hammer*. Simply possessing a tool orients us toward the world in a certain way (we look for nails to hammer), and encourages us toward certain ends (we aim to drive nails). This phenomenon is known, in the philosophy of technology, as tools' tendencies to *script* their users' behavior. And this phenomenon—the influence of everyday tools over how we approach and act in the world—operates not only at the level of individuals, but also at the level of entire societies.

To illustrate the effect our tools already have on us and our societies, consider an example from Langdon Winner. Winner shows how changing the design of one tool commonly possessed by a whole society—a power station—can change the attitudes and aims of the society and its individuals.[5] A nuclear power station, because it runs on easily weaponized uranium, encourages the group to see the world as containing hidden enemies who might threaten the society's security by stealing nuclear fuel. It further encourages the society to see certain measures of control and surveillance as reasonable, and encourages the society to pursue them. It is easy to see, however, that this particular orientation toward the world, and this particular set of aims would change if the design of the technology were changed from nuclear to wind or solar. Since these fuel sources are not weaponizable, individuals would see outsiders less as potential security threats, and the group would shift some of their energy away from aims of security and surveillance.

The result of this relationship between tools, individuals, and societies is this: As new tools are developed, new activities and ends are made possible and encouraged by them, and as we begin to value and pursue the ends to which our tools aim, the shape of "normal" human life changes, both at the individual and social scales. New kinds of jobs, social stations, habits, anxieties, institutions, and intentions—in short, entirely new ways of living human life—arise as a consequence of the technologies we adopt. Thus, what we value and what it means to live a characteristically human life are always being shaped by the technological world. This exposes the final flaw in the second assumption of the negative argument: the assumption that our normal, everyday technologies are passive tools we have total control over.

However, this raises an important question: How much power does the technological world have to shape our values? While some may argue that technological changes can amount to an entire refashioning of morality,

[5] Winner, Langdon. "Do Artifacts Have Politics?" in *The Whale and the Reactor*.

I am only making the much gentler claim that much of any group's view of the way human life should go is nuanced by (among other things) their particular socio-technological world. One important way in which this relation plays out deserves special mention here: the relationship between moral rules and the technological world.

Moral Rules and the Technological Background

Any given moral rule only makes sense, and can only be properly followed, in particular settings. The rule *one ought to/ought not to [insert some behavior]* only makes sense in a setting where the objects and actions involved in that behavior exist. Further, these rules only hold in worlds constructed so that the behavior has a particular (good or bad) effect or significance. Because the world that we inhabit is thoroughly populated by tools and tool-using customs, whether a particular behavior is possible and what *it* means or does will often be dependent on the particularities of our tool world.

To illustrate: We may hold that morality requires that we not cause needless harm to others, but in order to know *how* to follow that rule, we must determine which harms are necessary or unnecessary, and to do that we need to know a lot about the particular material, social, and tool world in which we are trying to follow the rule. If a time-traveling surgeon were to randomly teleport to some place and time where a surgery needed to be performed, how would she know how to follow the rule to avoid unnecessary harm? She would need to know what technologies were available in that world. Do they have anesthetics and antibiotics? Do they have robotic surgeons that can perform the surgery flawlessly, or do they only have scalpels, tourniquets, and liquor? What the moral rule—to cause no unnecessary harm to the patient—means, and what the physician must do to follow it, will be determined in part by the technologies that exist. This is true, generally; moral rules derive their meaning, and become applicable, only in relation to the specifics of the world, and this includes the technological situation.

Beyond the Negative Argument

We began with a sensible argument against the transhuman project. That argument objected to transhumanism on the basis that transhuman technologies, unlike others, meddled with human qualities that are fundamental to our very nature (primarily, qualities of the body and its functionality). Upon close reflection, however, we find that human nature and technology have always been mingled, and that their mingling has resulted in the constant transformation of the most basic dimensions of human nature. If we object to transhumanism for these reasons, we should therefore object to technology *in general*. So, unless we are prepared to embrace total Luddism, we must leave aside this argument against transhumanism.

Summary

This section aimed to describe the relation between humanity and technology in a way that would move us beyond the intuitive (i.e., immediately appealing), but mistaken, argument against transhumanism, and help us find a new orientation toward the transhuman program, both as humans and as Christians. We have arrived at the following position: Because human nature fundamentally involves our ability and tendency to create and use tools, the character of the normal and good human life will be partially determined by the nature of the particular world of technologies in which we live. Because human nature is inextricably connected with the creation and use of tools, and because the inevitable changes in the technological world translate into changes in how we live and what we value, the process of transforming ourselves and the world of our values is itself part of what it means to be characteristically human. Humanity is always engaged in profound tinkering with ways of living and being, and this means that the *general* project of transhumanism is not really a departure from our "normal" involvement with technologies. If there is an objection to be made to the transhuman program, it should be aimed at particular technological proposals, and the effects it will have.

The Future of Humanity

We have now left behind some of our initial negative intuitions regarding the possible transhuman future, but we still have reason to be skeptical about the utopian picture of the future painted by some transhumanist evangelists. In this section, I want to offer some reasons for adopting an attitude of hopeful agnosticism about whether a transhuman future would be a good or bad thing. The reasons have little to do with any of the facts about what transhuman technologies are being developed; rather they have to do with how the concept of a good or bad human life shifts in response to profound changes in the human world, and how we can admit the malleability of our concepts of good and bad without slipping into nihilistic relativism.

First, in order to get a clear view of things, it will help to wrap our heads around a concept called *baseline shift*.

Value Baselines and Baseline Shift

We acquire our sense of a normal and good human life through the process of orienting ourselves to the world into which we are born. Let us call this process *value calibration*. Through our experience we assemble, by collage, a complex image of what is valuable and important, and how life should be with reference to the things we value—things like private time spent with friends, the preservation of natural beauty, flourishing of economic markets, musical experience, the well-being of others, time with family, national pride, and thousands of other things we care about. We need a name for this complex, collaged image of how the world and our lives should be; let us call it our *value baseline*.[6] This baseline forms the ground from which our evaluative, ethical, and moral judgments are made.

Two features of the value baseline deserve attention. First, even though our value baseline is calibrated with reference to the specific historical moment we are born into, we tend to feel as if those values and expectations are not historically and culturally contingent, but ahistorical and objective. Second, because our historically conditioned baseline is what we use to register and evaluate changes, we tend to feel changes as losses or gains only if they happen *after* our baselines are calibrated; changes that happened before carry very little felt significance.

To illustrate: Those of us born before the development of the Internet and its various electronic devices tend to view these developments as resulting in a loss of interpersonal closeness. Those born after the Internet revolution, however, have calibrated their baseline expectations for how relationships operate *after* the shift. For this reason, they see technological mediation as a normal part of how relationships work, and view interpersonal closeness as involving technological mediation. Thus they may even view the Internet revolution as causing an increase in interpersonal closeness.

[6] I am adapting a term coined first by architectural theorist Ian McHarg, and later popularized by the ocean ecologist Daniel Pauly. McHarg, *Design with Nature*; Pauly, "Anecdotes and the Shifting Baseline Syndrome of Fisheries."

This leads to a quandary: when the older generation views a change as bad (i.e., a loss of something valuable) and the younger generation views the same change as good (i.e., a gain of something valuable), who is right? I claim that neither is any more right than the other, because each is relying on a different, valid conception of (in our example) interpersonal closeness, necessarily defined against the background of their technological world. To drive the nail just a bit further, consider that those of the older generation likely view the telephone as an aid to interpersonal communication rather than a hindrance to it. This, however, was not the common view of the pre-telephone generation.

The general point here is that every technological change involves the creation of new capacities that compete with familiar, valued ways of living. For this reason, almost any technological change can be framed as either a loss or a gain. Which way the change is seen is largely determined by which value baseline we view it from.

When technological change supplants familiar ways of being human and introduces new capacities, our baseline conceptions of a normal and good human life themselves shift in response to those changes, and a pattern emerges: Those living at the beginning of a significant technological change will often see it as a loss of something normal to human living, but those born after the shift will recognize their situation as normal rather than as a state of loss, and will view the technological change of the past as a natural development toward what human life *should* be like. Let us call this phenomenon *baseline shift*.

Since baseline shift recalibrates values each generation or so, different generations take different views of the goodness or badness of important changes. It is always difficult to make a judgment about which view is right, and both may be equally justified since each side is often grounded in its own coherent conception of the good human life. For this reason, there are rational viewpoints of the past that would view our world as a dystopic horror. Perhaps there is an imaginable world that every person from every viewpoint would call a dystopic horror, but such a unique world deserves its own name. Let us simply call it *Hell*. For any other case, dystopias do not exist independent of any viewpoint; rather, they are like a rainbow, which exists only in the view from a distance. This is worth remembering when we are horrified by visions of the transhuman future.

This raises a difficult question: How do we judge between those claiming that some possible future would be good and others claiming that it would be terrible? First, I suggest that these are not competing *factual* claims, but different value-laden *evaluations* of the situation. Facts and moral evaluations are not the same thing. This claim tends to make us bristle, because it seems uncomfortably close to saying that our deepest ways of valuing our lives and our world are a matter of preference. And *this* makes us uncomfortable because matters of preference are too often associated with *trivial* matters of preference—for example, whether one prefers chocolate or vanilla ice cream. "Certainly our deepest cares about the world are not akin to petty matters of taste!" we say. And we are right, but not because our values are *not* matters of preference, but rather because not all matters of preference are *trivial*. Whether we *prefer* a world where we all have privacy, where the rights of children are protected, where people of all races have equal legal rights—these are all preferences, but they are far from trivial; they are preferences *worth fighting for*. Perhaps our value judgments have a more fundamental status than this, but they do not need to in order to act as the foundation for our lives.

Morality: Referential, Not Relative

The picture I have drawn presents our values—even our deepest moral values—as at least partially products of our historical, social, and material conditions, since values, by their very nature, always relate and refer to the specific world we live in. Values that were acultural and ahistorical would, by definition, be pure abstractions without any content, and would have no practical relevance to our lives.

Some may worry that this skirts too close to the dreaded Moral Relativism that is rumored to be taught in The Secular University. But I am saying nothing of the sort. Even though there is a sense in which all these rules, norms, and values are dependent on (i.e., *relative to*) the material, historical, cultural, technological context we live in, it simply does not follow that there is nothing firm, grounded, or substantial to our values and moral principles. The fact that values and moral norms *refer* to the world does not imply that they are purely relative.

How Values Make Reference to the Technological World

The sort of referential way in which our values and moral norms depend on the world for their meaning can be illustrated perfectly by an analogy to the world of sports. In sports there are values and rules—like fairness, and injunctions to play fairly—which are akin to the values and moral rules of human life. What it means to play fairly and be a good sportsperson varies from game to game, and the proper rules of fair play can only be understood with reference to the material setup of a specific game. There is probably no nontrivial meaning of the notion of "fairness" that is common to golf, judo, volleyball, Greek wrestling, and bobsledding.[7] However, the fact that the values and rules of sports are all in a sense *relative* to the particular setup of each game does not therefore mean that the values and rules of sports are all "just relative," and therefore anything goes.

The relationship between the values of a good human life and the technological context in which those lives are lived parallels the relation between the values of sports play and the material context of the game: values and moral rules only have any instructive, meaningful content when considered against the backdrop of a specific material (technological) world.

Summary

We have now laid the theoretical, conceptual foundation necessary to move forward in adopting an informed attitude toward transhumanism. We have seen that the human relationship to technology is part of our very nature, and that our deepest values are themselves responsive to the world and the technologies that we use. And we see that this is not a bad thing, but merely the nature of values themselves. In order to move ahead in evaluating the transhuman project and the futures that it may bring, we must situate it not only within our new theoretical understanding, but also within the history of human-technology relations.

Three Familiar Transhuman Technologies

It is difficult for us today to understand the serious cultural anxiety over how now familiar technologies would fundamentally undermine what was seen as normal and good qualities of human life. Let us consider three examples of such humanity-transforming technologies: the printing press, the television, and the radio.

[7] By "nontrivial" I mean a notion of fairness that makes contact with the specifics of how any of the games are actually played. One could, of course, point out that all these sports share a common understanding of fairness as "sticking to the rules," but this trivial sort of definition only serves to illustrate the point I am making: that the rules and procedures of gameplay are so distinct that no definition of "fairness" that applies to all these games will provide any useful knowledge about how to *actually* play any of the games fairly.

The Printing Press as Transhuman Technology

Consider the printing press. As printing shops proliferated during the fourteenth century, so did anxiety over how the press would reshape normal and good human living. Chief among those anxieties was the idea that ubiquitous access to the printed book would undermine what was seen as a uniquely and fundamentally human faculty: our ability to store and access knowledge using purely internal, mental processes. If this new invention would allow anyone to store knowledge on the page and access it later, then it seemed inevitable that we would cease to use our minds in this way.[8] This was not merely the worry that we would be less educated or "smart"; it was the much more profound worry that the printing press would undermine the sort of mental lives that distinguish us from the lower animals: mental lives characterized by language-based knowledge and thought (which involved the combination of linguistic knowledge to produce new pieces of linguistic knowledge). In this way, the printing press genuinely threatened to destroy normal and good human living—now that any information could be printed, we would cease the characteristically human act of storing that knowledge as sentences in our minds and become more like animals, whose heads were comparatively empty of characteristically human kinds of knowledge.

It is easy to sneer at a moment of historical anxiety like this. However, the effects of the printing press were very much in line with what these critics anticipated. As knowledge was collected, organized, indexed, and printed, the art of memorization disappeared from the world. Today, long-form memorization, and the memory systems that enabled it—systems like memory palaces and mnemonic encoding systems—have all but completely disappeared and have been replaced by printed and digital information storage. We have, in fact, largely ceased to know things *in the way that was once considered uniquely and characteristically human.*

If the original critics of the printing press could get a glimpse of the world today, they would not sigh with relief; they would see their predictions largely fulfilled. Those of us born into the post-press world have a differently adjusted baseline conception of the normal and good human life though, and so we take a different view of the printing press, we cherish books as forms of art and repositories of knowledge that would otherwise be lost to history. The important point, however, is that the printing

[8] For a representative case of this argument, see Conrad Gessner, *Mithridate. Mithridates.*

press did in fact undermine what was seen as a fundamentally important feature of human life—one that distinguished us (much in the way *sapiens* does) from lower animals—and replaced it with a new, technologically mediated way of being human.

The Radio as Transhuman Technology

The theme continued as newer information-sharing technologies were developed. When radio sets became affordable and available to the public, there was no shortage of negative predictions about how it would undermine good human living. Most centered on the worry that the radio would lure our attention away from valuable activities of a full and flourishing human life. One particular worry deserves special mention: the worry that the radio would lure us away from one of the highest characteristically human pleasures—*reading*. What was once a threat to humanity had become an essential human activity.

Once again, the radio had just the anticipated effect. In 1936, the music magazine *The Gramophone* reported that children had "developed the habit of dividing attention between the humdrum preparation of their school assignments (mostly reading and recording printed information) and the compelling excitement of the loudspeaker."[9] And once again, those whose baselines had been set before the technological change viewed the post-change world as one in which patterns of the normal and good human life were undermined, while those born after the change had a different view of what a normal and good life was like—one that allowed for our attention to be divided between print and radio.

The Television as Transhuman Technology

At this point, we know the pattern well enough to predict how things went when the television came on the scene. Media historian Ellen Wartella notes how "opponents voiced concerns that television would … result in the further vulgarization of American culture"[10] by "undermining the [normal] patterns of family living," which included our use of "radio,

[9] *Gramophone*.

[10] Ellen A. Wartella and Nancy Jennings, "Children and Computers: New Technology. Old Concerns."

reading, [and] conversation."[11] Again the critics were right, but again those born into the post-television world were able to purse lives of value under their adjusted conception of what a normal and good human life is. The baseline had shifted.

Summary

We have seen in this section how now familiar technologies were, in the past, agents of transformations of human nature and ways of being that were viewed as fundamental for living a good and normal human life. Having seen that our world is in many ways already the product of radical technological changes that *transformed* humanity and helped us *transcend* our natural limitations, we can now turn to the final question: What orientation should the Church take toward the technological movement(s) that go under the name transhumanism?

The Church and the Transhuman Future

Up to this point we have concerned ourselves with the general relationship between human values and technology. Now that we have laid that general foundation, we can turn to the question of how the transhuman program fares in light of the particular values of the Church.

Conservatism and the Baseline

It is not uncommon for Christians to see humanity's simple beginning in Eden as the model for how the world ought to be. Technological change is then easily viewed with suspicion, and seen as an "outside" or "secular" force that makes human existence unnecessarily complicated and confusing, and that distances us from our natural and good original state. The role of the Church, from this viewpoint, is to act as a conservative force that resists changes that alienate us from our own nature.

While this attitude is a common and intuitive one, the Church has not historically been conservative in quite this way. Rather, it has been conservative not of a larger set of trans-historical Christian values, but rather of the baseline values and expectations of its living members—a baseline that was calibrated in response to the world in which they were born. The

[11] Ibid.

result is often that, when some change begins, the Church is immediately resistant, but its resistance fades once the next generation, whose baselines were calibrated after the change, become its leadership.[12] At this point, the Church's resistant efforts shift toward whatever new change threatens the new baseline.[13] Let us call this phenomenon of initial resistance to changes that threaten current baselines *reactionary baseline conservatism*. This sort of conservatism is the easiest to adopt, but it poses the danger that each new generation will accept what the previous generation saw as drastic, while shifting their concern toward the newest change, which will be accepted easily by the next generations.

This picture of the Church's susceptibility to baseline shift may trouble some. Is the Church really so susceptible to changes in its core values? We must not overstate the situation here. Part of the nature of shifting baselines is that it is not always a core value *per se* that has shifted, but rather our understanding of *how that value is to be expressed or pursued*. This means that the core value remains intact, but its expression or meaning has shifted.[14] To return to an earlier example, it might seem that the rise of digital communication has threatened to undermine the value we place on interpersonal connections. In fact, what has shifted is not the value itself, but *how we pursue or express* that value. We once thought that closeness was best expressed through discrete face-to-face interactions, but many now instead see a constant connection-over-a-distance as the way to pursue interpersonal closeness. In this case, the value *itself* has remained intact even though a shift has taken place in our baseline conception of what that value *means*, and how it is to be *pursued*.

Problems with Reactionary Baseline Conservatism

This sort of simple, reactionary conservatism of the baseline poses three dangers to the Church. First, it makes it easy for the focus of the Church's efforts to shift every generation or so. With each shift of focus the Church

[12] I use the term "generation" here in something less than the strict, literal sense.

[13] It is worth noting that, for reasons to do with organizational structure, the Catholic Church has often fared better in maintaining a coherent conservative stance than the Protestant Churches have. This is, however, not to say that the conservative stance is a good one.

[14] We can also acknowledge that there are some "core" values that are intentionally upheld across many generations and are therefore somewhat more resistant to long-term change. These do not seem to include technology-oriented values.

inevitably concedes some of the ground it defended in the previous struggle. Secondly, because the Church's conservatism is usually protective of the baseline of its present members, the Church will often fight to maintain standards that, just a generation or two before, it fought against. Because each generation's baseline view of a normal and good human life is calibrated anew, the baseline ways of living that the Church fights to preserve are often the very things that formerly threatened, and replaced, cherished ways of living a normal and good human life. Finally, the most significant problem with the rationality behind this sort of conservatism is that it rests on a tendency to see threats to *our* valued ways of living a good human life as threats to living a good human life *at all*. This reasoning is the root of much wasted resistive effort on the Church's part, and runs counter to the spirit of hope and openness to new and good forms of human life which ought to characterize the Church.

The best way to counter the pernicious effects of naive, reactionary baseline conservatism is for the Church to explicitly recognize, in the face of approaching radical change, that good human lives can take a stunningly wide variety of forms in response to the social and material contexts in which they are lived. The ways of being a happy human are surprisingly malleable—so much, in fact, that the earliest Christians would certainly be totally shocked and disoriented by the modern world and the way the Church exists in it. Although it is easy to imagine the rational horror the Apostles may have had to telecasted church meetings, modern reordered marriage dynamics, or the use of Twitter, Facebook, and video games for religious ends, this alone does not give us reason to "go back" to any earlier way of doing things. Baselines have shifted, and with them so have conceptions of the forms good human and Christian lives can take. The one thing that will be consistent throughout the future is this pattern.

Conclusion

Like all past futures, whichever future the Church inherits will inevitably be a transhuman one. So we bear the following paradoxical relationship to the future: On the one hand, we are justified in seeing coming transhuman changes as dystopic, since they will inevitably supplant our conception of a normal and good human life. On the other hand, those born into the future world may be equally justified in not seeing their world as dystopic since this new state will constitute new standards by which genuinely normal and good human lives might be pursued. In fact, the transhuman future, when it comes, will seem no more radically different from our world than ours would seem to worlds past. There will certainly be radical changes, but I see no reason to conclude that those changes are a departure from the history of self-imposed technological change rather than an extension of the very sort of radical changes with which humanity has always been engaged. The changes may come more quickly, and they may involve new tools, but they are not therefore part of a different relationship between technology and human nature.

There is therefore no simple answer to the question posed in the beginning of this chapter. For its earliest, initial recipients, the transhuman future may be genuinely bad. It will undoubtedly undermine familiar, valued ways of pursuing good lives. However, for those born into that future, who have had their values and attitudes calibrated in the transhuman world, that future may provide the necessary ground from which new forms of genuinely good human life grow—forms of good human life that we would not recognize now.

It is impossible to take any view—positive or negative—toward transhumanism as a whole, since our entire history is in many ways already a transhuman one. It is also naive to reject change as bad simply because it may disrupt familiar ways of living. We know that what is good about being human is itself something that shifts drastically across history and culture, because new goods and forms of human well-being arise and supplant old ones. For this reason, we have good reason to think that our horror at the ways the future could go is not a good measure of how good or bad the future will be for those who actually inhabit it; neither is it a good measure of how well or poorly those people will be able to live lives that fulfill the Church's charge to honor God and love one another. For these reasons the Church must work not only to maintain a healthy skepticism toward technological changes, but also to maintain openness and generosity of spirit toward the (inevitably strange) new worlds that may emerge. Because we are situated outside those possible future worlds, we probably lack the understanding of their nuanced values and concepts necessary for making a judgment about whether that future is a good or bad one for them. Humanity has survived the emergence of radically new worlds already, and this is a testament to the versatility of human nature.

Consequently, humanity in general, and the Church in particular, should exercise caution in their appraisals of technological changes. We should even perhaps adopt an agnosticism about whether radical technological changes will hinder or aide our pursuit of good human and

Christian lives. When we have the power to affect the course of our technological future, we should aim to maximize the benefits we can clearly foresee, and minimize the harms. In addition, though, we should leave open in our minds a third category: consequences that seem strange, even disorienting to us, but which are not obviously beneficial or harmful. About these we should temper our concern with hope, remembering that humanity has remade itself many times over already, and in doing so has not only imperiled itself but also created myriad new forms of human flourishing.

We can see that whatever future comes, it *will* be a transhuman one, as all radically different futures have been. That future will supplant familiar and cherished ways of being human, even ones that are so familiar to us that they seem natural and inviolable. And it will replace our familiar forms of humanity with new ones that would seem completely alien and inhuman to us were we to see them through a crystal ball. However, humanity has already demonstrated profound malleability over the course of its brief existence, and this gives us good reason to think it will still be possible for there to exist societies, relationships, and individual human lives characterized by new forms of both joy and misery, things beautiful and hideous, and new forms of strife, love, and holiness.

Bibliography

Arendt, Hannah. 1958. *The Human Condition*. Chicago: The University of Chicago Press.

Becker, Ernest. 1976. *The Structure of Evil: An Essay on the Unification of the Science of Man*. New York: Free Press.

Berger, Peter, and Thomas Luckmann. 1966. *The Social Construction of Reality: A Treatise in the Sociology of Knowledge*. New York: Doubleday.

Bostrom, Nick. 2005. Transhumanist Values. *Journal of Philosophical Research* 30 (Supplement): 3–14.

Gessner, Conrad. 2009. In *Mithridate. Mithridates (1555)*, ed. Bernard Columbat and Manfred Peters, 14–48. Geneva: Librairie Droz S.A.

Gramophone, September 30, 1936.

Halton, Eugene. 1995. *Bereft of Reason: On the Decline of Social Thought and Prospects for Its Renewal*. Chicago: University of Chicago Press.

Heidegger, Martin. 2008. Letter on Humanism. In *Basic Writings*, ed. David Farrell Krell, 141–183. San Francisco: Harper Perennial.

Huizinga, Johan. 1949. *Homo Ludens: A Study of the Play-Element in Culture*. Boston: Routledge.

Jensen, Derrick. 2006. *Endgame, Volume II: Resistance*. New York: Seven Stories Press.

Kurzweil, Ray. 2009. *Transcendent Man*. Directed by Barry Ptolemy. Ptolemaic Productions and Therapy Studios.

McHarg, Ian L. 1995. *Design with Nature*. 1st ed. New York: Natural History.

Pauly, Daniel. 1995. Anecdotes and the Shifting Baseline Syndrome of Fisheries. *Trends in Ecology & Evolution* 10 (10): 430.

Tallis, Raymond. 2007. Enhancing Humanity. *Philosophy Now* 61 (May/June). https://philosophynow.org/issues/61/Enhancing_Humanity

Voland, Eckart, and Wulf Schiefenhövel, eds. 2009. *The Biological Evolution of Religious Mind and Behavior*. New York: Springer.

Wartella, Ellen A., and Nancy Jennings. 2000. Children and Computers: New Technology. Old Concerns. *The Future of Children* 10 (2): 31–43.

Winner, Langdon. 1989. *The Whale and the Reactor: A Search for Limits in an Age of High Technology*. Chicago: University of Chicago Press.

CHAPTER 11

Origin Stories: Superheroes, Cyborgs, Artificial Intelligences (and Other Humans and Posthumans)

Jeanine Thweatt

Origin Stories, Traumas, and Heroes

Let us begin with a consideration of the most familiar of posthuman figures: the superhero. These paradigmatic more-than-human characters may not be technologically enhanced in any obvious way; in fact, in their pedestrian, secret, merely human identities, they must not be, or they would be immediately recognizable and the narrative falls apart. Yet they are clearly posthuman in the broad sense of being characters conceived of as transcending ordinary human capacities, both physically and mentally.

You don't need to be a faithful attendee of Comic Con to know that all superheroes, and their requisite supervillain counterparts, have fantastic origin stories which transform them in some essential way, and henceforth (re)define their identity. Indeed, as (at least some) superhero narratives have moved out of strictly graphic novel and serial comic book territory to

J. Thweatt (✉)
Flagler College, St. Augustine, FL, USA
e-mail: jthweatt@flagler.edu

S. Donaldson, R. Cole-Turner (eds.), *Christian Perspectives on Transhumanism and the Church*, Palgrave Studies in the Future of Humanity and its Successors,
https://doi.org/10.1007/978-3-319-90323-1_11

be obsessively and lucratively told and retold in the sacred, dark forum of the American movie theater, these narratives have claimed a decisive presence in mainstream pop culture. No one here can fail to know who Spider-Man is, though you may not know that to properly spell his name you must use a hyphen.

I hardly need observe that, quite often, these origin stories depict superheroes as the unwitting recipients of technology gone awry. This is the genre's specific rearticulation of a literary trope traceable back to Mary Shelley's *Frankenstein*. Superheroes, like Frankenstein's creation, are instances of technologically transformed posthumans, whether by design or radioactive spider-bite or ingestion of experimental serum or terrible industrial accident or (insert your favorite origin story here). What also becomes exceedingly clear, upon reflection across the wide swath of superheroes in print and film, is the common dimension of trauma in the origin stories of superheroes: "These origin stories are, almost without exception, marked with incidents of trauma: the murder of Spider-Man's uncle, the death of Batman's parents, the destruction of Superman's home planet, etc."[1]

As Philip Sandifer observes, the operative past of a superhero at any given moment is best described as an accumulation of trauma, the most obvious of which is the trauma of the origin story, "which is typically the primary motivation for why the character dresses up in spandex and fights crime."[2] Yet, as he further observes, this initial origin trauma is far from the only one operative in the formation of superhero identity; the life of a superhero is rife with traumas, given the superhero's compulsion to combat crime, injustice, and tragedy at every opportunity. Yet, not all accumulated traumas are actively remembered in superhero narratives—some fade, while others become part of the active formation of identity going forward. Why, asks Sandifer, are some forgotten, while others are remembered? The difference, he suggests, is that the traumas remembered are those that mirror the original trauma in some fashion, and thus become an extension of it: "All of the important and recurring Spider-Man traumas, for example, revolve around Spider-Man's failure to save someone's life and guilt over that. All of Batman's revolve around anger at some external force or an act of violence—whether the killer of his parents or the Joker. … They do not resolve the origin trauma, but rather repeat it

[1] Sandifer, "Amazing Fantasies: Trauma, Affect, and Superheroes," 175.
[2] Ibid., 176.

so perfectly that they themselves become irresolvable traumas—at once repetitions of the origin and new origin points unto themselves."[3]

One must also observe that the origins of the superhero's narrative counterpart, the supervillain, can be described in similar terms. It becomes fruitful, then, to ask—what separates the supervillain from the superhero? What factor is it that turns one character to the superhuman task of righting wrongs, and another to the superhuman perpetration of wrongs? Of course, this is simply another way of asking what it is that turns each of us one way rather than another—as these stories, like all stories regardless of fantastical or futuristic settings, are simply ways of exploring what it means to be human in all of its tragic, moral, and immoral dimensions.

The answer I will suggest here is simple: what marks the difference between superhero and supervillain is how they live out their transformed identity in the aftermath of their origin trauma. Superheroes wrest from the trauma, never resolved and always ongoing, a determination to devote their superior powers to the common good in some way—a reaction that suggests that, at some level, they continue to identify as human despite the transformation into superhuman hero. And therefore, of course, what makes a villain is the opposite: in the aftermath of the origin trauma, supervillains lose all sense of continuity with humanity, including their own human identity prior to the trauma, with the result that humans no longer merit moral status or regard.[4] In this context, Sandifer's observation—that it is the superhero identity that is primary—takes on a dimension of special importance. "Because the organizing pathology of the superhero comic is that of a post-traumatic identity, we are obliged to read any claim to a 'prior' identity as a construct of that post-traumatic identity ... it is not accurate to say that Peter Parker became Spider-Man, but rather that Spider-Man created Peter Parker to have become him."[5] It is this act of post-traumatic identity construction, struggling to maintain a meaningful continuity with human identity, which marks the difference between hero and villain.

What does this brief exploration of the origin stories of superheroes and supervillains—completely fictional characters inhabiting impossibly fantastic narrative worlds—tell us? After all, we want to be talking about

[3] Ibid., 178.

[4] This simple dichotomy can be complicated by further analysis of particular superhero figures; for example, see the essays on Superman and Batman in Oropeza (editor), *The Gospel According to Superheroes.*

[5] Sandifer, "Amazing Fantasies: Trauma, Affect, and Superheroes," 182.

current reality and possible futures. These splashy tales written in broad strokes on a huge canvas, these caricatures of ourselves, give us an oversized reflection of the truths of our own smaller, mundane lives. We too have our origin stories which shape our notions of who we are and what we must do. And it is my suspicion that the great insight of the comic book superhero genre is that all origin stories are marked inevitably by trauma.

From Superhuman to Posthuman

At its most fundamental, the question we confront in reflecting upon the notion of the posthuman is a question of continuity and discontinuity with the human, as (we think) we know it. The greater the perceived, or assumed, discontinuity, the more frightening the posthuman future becomes. This is where the posthuman becomes monstrous, as Elaine Graham so aptly elucidates in her interpretation of everyone's favorite bald Star Trek Captain, Jean-Luc Picard, and his transformation into the horrific Borg Locutus.[6] Conversely, the more closely particular visions of the posthuman align with our sense of "the human," the more benign, reasonable, warranted, and indeed desirable the posthuman begins to appear.[7]

But what, after all, do we mean by "continuity with the human?" This must surely go beyond crude similarity of form—though of course the trope of signifying evil with physical ugliness is as old as fairy tales. It is tempting, perhaps, to reach for continuity of function as an alternative; perhaps a posthuman that goes only so far, but no further, beyond present human physical and mental capacities is morally justifiable and intellectually acceptable. Yet this strategy falls afoul of the basic problem present in all bioethical debates regarding the line between "therapy" and "enhancement"—namely, that the line is fuzzy, shifty, ultimately arbitrary, and highly relative.

In this context, the superhero's determination to construct a post-traumatic, posthuman identity that maintains significant social and moral connections to human identity is instructive in how to navigate notions of potential posthuman identities in possible futures. Here, continuity is not crudely morphological nor is based in some cautious restriction around what sort of enhancement is acceptable in the posthuman; rather, the

[6] Graham, *Representations of the Posthuman*, 132–152.
[7] Thweatt-Bates, *Cyborg Selves*, 2.

superhero's (the posthuman's) continuity with humanity is a matter of the hero's maintaining and continuing to cultivate meaningful relationships with the humans around him or her, even when this means a continual painful consciousness of difference and even a sense of some degree of personal isolation. The question we ought to be asking, I suggest, is not whether we (humans) consider a particular potential posthuman identity human—but whether that posthuman does.

Humanity's Children and Childhood's End

This is, of course, simply one way to redescribe the basic anxiety around which a myriad of science fiction narratives revolve.[8] These narratives exist because the anxiety is real. Shortly before writing this chapter, an article popped up on my Facebook news feed, posted by an acquaintance who teaches ethics at a nearby seminary, about roboticist David Hanson's conversational Philip K. Dick android, who apparently sought to reassure his PBS NOVA interviewer that, when he and the other robots took over the world, "[b]ut you're my friend, and I'll remember my friends, and I'll be good to you. So don't worry, even if I evolve into Terminator, I'll still be nice to you. I'll keep you warm and safe in my people zoo, where I can watch you for ol' times' sake."[9] The amount of alarm expressed by various friends and friends of friends in the conversation thread that followed seemed to me to be quite overblown. Cutting-edge as this android and its programming is, it does not yet possess an active sense of irony, much less genocidal hostility. Yet the worry is hard to shake and not without validity.

The classic Arthur C. Clarke tale, *Childhood's End*, tells the story of the sudden, *en masse* evolution of a generation of humanity into a powerful, disembodied collective entity that has no capacity to relate to the human generation that gave it birth. The erosion of the intimacy of human parent and posthuman child is perhaps the most powerful narrative in the science fiction genre of the anxiety surrounding the nature of the posthuman and the question of its relation to the human.

[8] For a full discussion of dystopian science fiction narratives in dialogue with the contrasting optimism of transhumanism, see Dinello, *Technophobia!*

[9] Draper, "AI robot that learns new words in real-time tells human creators it will keep them in a 'people zoo.'"

This same anxiety is retold in the recent Syfy reboot of *Battlestar Galactica*. Consider the contrast between the Cylon characters of Caprica Six and Brother Cavil. Again and again, across the entirety of the series, both Six and Cavil refer to the Cylons as "humanity's children"—and neither means this as a compliment. Caprica Six uses the identity as an explanation for the Cylon's retributive genocide, implying that to kill is fundamentally part of human nature; Cavil chafes against this identity in his desire to "become the best machines we can be" as opposed to the others' futile and humiliating impulse to ape flesh-and-blood humanity. Cavil's ideological refusal to identify with humanity, his distaste for the biological body, and his impatience with its limitations make him the ultimate posthuman villain. He is the artificial intelligence (AI) equivalent of the bug-eyed human-eating alien monster: an enemy with nothing in common with us and who cannot be reasoned with.[10]

But let me draw us back to the "origin story" that launches *Battlestar Galactica* to begin with, the origin story that appears at the frame of every episode: "The Cylons were created by Man. They rebelled. They evolved. They look and feel human. Some are programmed to think they are human. There are many copies. They have a plan."[11] This story, which informs the Cylon construction of identity, is a story of trauma, one which issues in an imperative to self-protect by annihilation of the other—and yet also contains an acknowledgment of humanity as creator/parent, and unfolds in the desire to become more human rather than less, resulting in the Cylon evolution of "skinjobs" (i.e., realistic human bodies). This conflicted origin story, in which the creators, humans, are also the killers of their creations, in which the "children of humanity" must "kill their parents in order to come into their own" is never given in full detail. It functions like Sandifer's description of the superhero origin story: constantly, obsessively rehearsed in a sort of liturgical mode, in symbolic shorthand, and reenacted into fresh trauma. This is, simply, what origin stories are: condensed, symbolic statements of who we are and what we must do.

[10] And yet it is clear that Cavil's hatred of humanity—and his own concomitant self-hatred for his biologically human form—is no ontological necessity, for Caprica Six and the other sympathetic Cylon models, despite their distrust of humanity and the history of genocidal violence, deeply desire to identify as human, paradigmatically explored in the narrative in the most intimate of relationships.

[11] Larson and Moore, *Battlestar Galactica*.

(Post)Humans in the Garden

So what, then, does our own Christian origin story tell us about who we are, and what we must do? What is the trauma that marks the origin of humanity, and what is our response to it in the Christian creation story?

The story begins, as we know, with a description of perfect intimacy: the first humans walk and talk with their creator in the cool of evening—a scene forever inscribed in my memory through the hymn lyrics "and he walks with me, and he talks with me, and he tells me I am his own."[12] And yet this perfection does not last; it is broken through an act of humanity, which can be read as either defiance or a desire for greater intimacy (or, perhaps these need not be mutually exclusive). And herein enters the trauma of our Christian origin story, the event that subsequently defines human identity: we are expelled from the Garden, that place of perfection and intimacy with the Creator, and live out our now mortal lives in a realm of pain and suffering, toil and sweat.

But the story goes on; what it describes is the struggle of humanity to continue in relationship, however imperfect, with the Creator God to whom it owes existence and identity. And the crux of this struggle is the question of how to react to this origin trauma: do we rebel against the Creator who punished us, or do we continue to reach for a relationship of intimacy with this God who created us?

This question plays out across the narratives of the Hebrew Bible, the drifts away from the Creator and the repeated returns in response to the steadfast love of the Creator. And, for Christians, it culminates in the incarnation of God in the person of Jesus, the Christ, whose work is the reconciliation of the human to the divine.

Children of God

The question, then, for the posthuman is no different than the question of the human, in relation to its origin: what is the relationship with the creator? For transhumanists, this question plays out as a real-world problem in the quest to construct "friendly AI." This quest requires proactive problem solving on the front end, before our posthuman progeny become sophisticated enough to unironically and intentionally attempt to reassure us with prospects of people zoos. Nick Bostrom and Eliezer Yudkowsky,

[12] Miles, "In the Garden."

among others, have thought through, and written about, potential catastrophic AI scenarios and offered what they believe is our best hope of avoiding them. Yudkowsky, in particular, offers the observation that our only hope of avoiding apocalyptic AI scenarios is to make sure that whatever AI we construct doesn't *want* to destroy us, because, he posits, it will certainly be able to. The problem lies in *how* to accomplish this. At this point, Asimov's Three Laws seem quite hopelessly dated—in any case, as he himself and any reader of his work are quite aware, the ability to subvert those not-quite-airtight laws is what every robot narrative of his corpus turns on.

I believe what Yudkowsky is intuiting, correctly, is that no attempt at coercion, no matter how thorough and proactive, is likely to be successful. Yudkowsky believes this is so because of the godlike nature of self-enhancing AI, which will so outstrip us that not only will there be no hope of controlling it, but also no hope that it will not recognize its own superiority over its creators. And here we find a significant divergence between the origin story of humanity and this version of the posthuman: the created supersedes the creator, and the underlying metaphor of Western origin stories of the father-creator and the created child is reversed. This is the divergence which generates the anxiety around the human–posthuman relationship. Will our posthuman creations be rebellious children? Worse still, will they be resentful, or even simply indifferent gods?

Correct though I think Yudkowsky's pessimism is at controlling AI, I do not believe the difficulty is best attributed to the godlike powers he assigns to such a posthuman creation. I think, rather, that the problem lies in the antagonism that such preemptive control presupposes, and thus creates, as a self-fulfilling prophecy. One need not be godlike to resent coercion—not if my attempts to put my foot down with my 9-year-old daughter about tooth-brushing have taught me anything at all. The problem is much more fundamentally *relational.*

Mother-Creator

This suggests that the answer to the "friendly AI" problem, and the larger question of the posthuman's identification with the human altogether, is a matter of achieving right relationship. This, in turn, leads us back to the primary metaphor of parent and child, operative not only in the theological context of the Christian origin story, but more generally (and not coincidentally) in the Western context; specifically, for our purposes, in the genre of science fiction. If the horror scenario is enshrined in narratives of the births of monsters and the alienation of child to parent, then the ideal scenario is one in which the relationship of parent and child is strong and unbroken.

The insight which Christian theology brings to this issue is basic and simple, and based in the story of steadfast love represented, over and over, in the biblical narratives which depict God as initiating and reinitiating relationship with humanity, despite all failures and betrayals.[13] This is a model of parental love that might well be extended, not simply to our relationship with our human children, but our anticipated posthuman children as well, in whatever form they may take. Importantly, such a model precludes the kind of preemptive coercion that dooms itself to failure, and presents, instead, a relationship that honors the integrity of the other, even if that necessarily demands the acceptance of the possibility of repudiation.

As a practical matter, this means that the task of spiritual formation, one of the most fundamental tasks of the Christian community and the church, must focus on grounding our relationship to others, human and nonhuman, within the theological context of this unbroken, parental, steadfast love. This is, of course, not a new insight, but perhaps articulating the necessity of extending this model of relationality beyond the neighbor, beyond, even loving our (human) enemy, into relationships with others that we have yet to encounter or imagine can give it fresh force. How do we teach our children, who will undoubtedly be the creators of technological marvels we ourselves cannot envision, to relate to these creations in ways that respect their integrity? How do we teach our children that others are not automatically to be, at best, exploited to our own advantage and, at worst, feared and annihilated?

As a brief attempt at answering this I return to the basic function of science fiction narratives, which ultimately is not simply to describe the imagined future but to simultaneously critique the present. These narratives are instructive not only as warnings of potential future missteps but as descriptions of current moral failures. Cylons and other posthuman figures are stand-ins for the maligned, oppressed, and hunted members of

[13] I use "Mother" in the heading deliberately in recognition of the unfortunate distortions toward hierarchy, absolute and arbitrary authority, and relational distance that the role and relationship of "Father" has suffered within Western culture, philosophy, and theology—though by doing so I do not mean to endorse a gendered dualism of masculine and feminine natures.

humanity that we even now deny full moral consideration and status.[14] If we are to teach our children how to relate to their potential posthuman creations, we must take to heart the lessons from the prophets—both of our times and from our scripture—and apply them to those who even now are not considered fully human.

A final note: this may in fact mean that as we move toward the creation of the posthuman, we are indeed "playing God" in a real and very serious way—not in the sense that we are somehow usurping the role of our own Creator, but in the same sense that the creation of any child is "playing God" and implies responsibility for that creation. In the words of William Adama, Commander of the *Battlestar Galactica*, in an impromptu speech that is as much a confession of parental failure to his listening son as it is a comment on humanity's failure in relation to the Cylons:

> When we fought the Cylons, we did it to save ourselves from extinction. But we never answered the question why—why are we, as a people, worth saving? We still commit murder, because of greed, spite, jealousy. And we still visit all of our sins upon our children. We refuse to accept responsibility for anything that we've done. Like we did with the Cylons. We decided to play god; create life. When that life turned against us, we comforted ourselves in the knowledge that it wasn't really our fault, not really. You cannot play god, then wash your hands of the things you've created. ... Sooner or later, the day comes when you can't hide from the things that you've done anymore.[15]

Will our posthuman children recognize themselves in us? Will we recognize ourselves in them? Will we accept the creator's responsibility for our creations?

Bibliography

Dinello, Daniel. 2005. *Technophobia! Science Fiction Visions of Posthuman Technology*. Austin: University of Texas Press.

Draper, Chris. 2015. AI robot that learns new words in real-time tells human creators it will keep them in a 'people zoo'. *Glitch.news*, August 27. http://glitch.news/2015-08-27-ai-robot-that-learns-new-words-in-real-time-tells-human-creators-it-will-keep-them-in-a-people-zoo.html

[14] I am thinking here of both the current dehumanizing rhetoric in American politics around Islam and the realities of the carceral state which imprisons a disproportionate number of African Americans.

[15] Larson, *Battlestar Galactica* (original miniseries).

Graham, Elaine. 2002. *Representations of the Posthuman: Monsters, Aliens and Others in Popular Culture*. New Brunswick: Rutgers University Press.

Larson, Glen A. (creator). 1978–1979. *Battlestar Galactica*. TV Series.

Larson, Glen A., and Ronald D. Moore (creators). 2004–2009. *Battlestar Galactica*. TV Series.

Miles, C. Austin. 1912. In the Garden. *Hymnary.org*. https://hymnary.org/text/i_come_to_the_garden_alone

Oropeza, B.J., ed. 2005. *The Gospel According to Superheroes: Religion and Popular Culture*. New York: Peter Lang.

Sandifer, Philip. 2008. Amazing Fantasies: Trauma, Affect, and Superheroes. *English Language Notes* 46 (2): 175–192.

Thweatt-Bates, Jeanine. 2012. *Cyborg Selves: A Theological Anthropology of the Posthuman*. Farnham: Ashgate Publishing.

CHAPTER 12

Even Cyborgs Cast a Shadow: Christian Resources and Responsibilities in Response to Transhumanism

Fred Glennon

Introduction

Many critics of transhumanism focus solely on the negative or "shadow side" of transhumanism's attempt to transcend our creaturely limits by exploring the potential social and ethical disasters awaiting any unimpeded technological march toward the transhumanist utopian vision of immortality. Of course, the word shadow has numerous meanings. Let me highlight two. First, shadow can mean an imperfect or feint representation of something. This is what some of transhumanism's critics are doing when they engage select transhumanists such as Ray Kurzweil, and ignore the diversity of perspectives represented in the transhumanist movement, many of which share the same liberal, humanistic values the critics hold. (Yet, as this chapter will demonstrate, even these liberal, humanistic accounts of transhumanism have their shadow, especially in their imperfect

F. Glennon (✉)
Department of Religious Studies, Le Moyne College, Syracuse, NY, USA
e-mail: glennon@lemoyne.edu

S. Donaldson, R. Cole-Turner (eds.), *Christian Perspectives on Transhumanism and the Church*, Palgrave Studies in the Future of Humanity and its Successors,
https://doi.org/10.1007/978-3-319-90323-1_12

representations of the self and political community.) A second meaning of shadow is as a constant companion or follower. The song "Me and My Shadow" reflects this meaning. We see this aspect of shadow among a number of theologians and ethicists who have been following the transhumanist movement for years, seeking connections between transhumanist aspirations and Christian theological and ethical concepts. They have brought light to the shadows of narrow understandings of transhumanism and challenge Christians of all stripes to engage more fully rather than dismiss the insights and challenges reflected in the transhumanist movement.

In this chapter, I will build upon the metaphor of shadow as both imperfect representation and constant companion to highlight the strengths and limits of transhumanist visions of the posthuman self and political community and to call the church to respond in meaningful ways to the challenge the movement represents. In the first section, I discuss briefly two images of the posthuman: the individualistic view of transhumanists and the relational conception of cyborg of some theologians. Both recognize the need to employ human enhancement technologies responsibly, but the relational view of the cyborg reflects better, although not completely, the social nature of responsible selfhood as articulated by H. Richard Niebuhr. Niebuhr's emphasis on social accountability, so necessary to provide the needed restraints to hold in check the negative potential of human enhancement technologies, eliminates the shadows of the other views of the self. In the second section, I contend that the conception of political community which underlies the transhumanist movement is too narrow and too imperfect to guarantee that the human enhancement technologies will be developed and employed justly and equitably. As an alternative, I draw upon the Christian tradition of covenant community, which reveals more fully our social selves in community and provides a stronger basis for the role of the state in securing the common good. Finally, I call upon the Church to follow the path of a number of theologians and ethicists to shadow the transhumanist movement, both to draw attention to the ways it can deepen Christian theology and to shed light on the potential dark side such technological enhancements can pose for humanity and the world.

A Shadow of the Responsible Self

What is the transhumanist vision of the self? The answer is found in this definition of transhumanism by Nick Bostrom:

> Transhumanism is a loosely defined movement that has developed gradually over the past two decades and can be viewed as an outgrowth of secular humanism and the Enlightenment. It holds that current human nature is improvable through the use of applied science and other rational methods, which may make it possible to increase human health span, extend our intellectual and physical capacities, and give us increased control over our own mental states and moods.[1]

The result of these efforts to improve human nature may be substantively superior "posthuman" beings who have exceptional intelligence, longer life spans, and greater moral virtue.

Key for purposes of this chapter is the understanding that this vision of the posthuman is a continuation of the humanism of the Enlightenment with its conception of the self as autonomous and free, unencumbered by its relations with others. The Enlightenment emphasized the need to incorporate reason, autonomy, and progress into all spheres of life to overcome the limits placed upon humanity by nature and superstition (read Christian religion). Immanuel Kant argues for self-rule in the realm of morality because the self, through its use of reason, can discern and abide by universal moral standards. Adam Smith advocates self-rule in the marketplace—individuals rationally pursuing their own goods will ultimately result in an economically prosperous nation. Transhumanists want self-rule over human biology, which Simon Young calls morphological freedom: "As humanism freed us from the chains of superstition, let transhumanism free us from our biological chains."[2] All three affirm the Enlightenment vision of using the human mind's ability to reason as the means to generate such autonomy. All three articulate a kind of universal framework that transcends cultural limits, and, with transhumanism, transcends biology.

The individualistic nature of such morphological freedom is evident among many transhumanists. Max More argues: "True transhumanism does seek to enable each of us to alter and improve (by our own standards)

[1] Bostrom, "In Defense of Posthuman Dignity," 202.

[2] Young, *Designer Evolution*, 32.

the human body."[3] Nick Bostrom affirms the value of individual freedom and choice when it comes to the use of enhancement technologies as long as they do no harm to someone else. In a statement reminiscent of the argument of John Stuart Mill about freedom over one's own person in *On Liberty*, Bostrom writes:

> People should have the right to choose which enhancement technologies, if any, they want to use. In cases where individual choices impact substantially on other people, this general principle may need to be restricted, but the mere fact that somebody may be disgusted or morally affronted by somebody else's using technology to modify herself would not normally be a legitimate ground for coercive interference.[4]

To their credit most transhumanists do contend that we must use such technologies responsibly. Nick Bostrom argues that "Transhumanists place a high value on improvements in our individual and collective powers of understanding and in our ability to implement responsible decisions."[5] Not to use these technologies, they suggest, would be irresponsible and may result in the end of civilization as we know it.

While many Christian theologians and ethicists applaud the desire on the part of these transhumanists to develop and employ these technologies in a responsible manner, they are not sure such other-regarding responsibility can be grounded well in their individualistic, socially (and biologically) unencumbered conception of the self.[6] Just as Christian ethicists and theologians have been critical of the liberal humanist view of the self, they are critical of the way in which that vision of the human gets extrapolated into the transhumanist vision of the posthuman. Ron Cole-Turner suggests that, from a Christian point of view, "transhumanists seem both human-centered and self-centered, concerned chiefly with

[3] More, Max, "True Transhumanism: A Reply to Don Ihde," in Hansell and Grassie, *H+/-: Transhumanism and Its Critics*, 143.

[4] Bostrom, "Transhumanist Values," 11.

[5] Ibid., 12.

[6] Thweatt-Bates questions the transhumanist view of the human self as a disembodied center of consciousness and will that uses technology but is unaffected by it: "It is hard to miss the parallel between Neo-platonic Christian theological views of the body as 'evil, seductive matter' and the transhumanist view of human biological bodies as placing negative limits on human potential." J. Jeanine Thweatt-Bates, "Artificial Wombs and Cyborg Births," in Cole-Turner, *Transhumanism and Transcendence*, 109.

humanity above other species and with themselves among other humans."[7] Christians recognize that such self-centeredness is a failing, not a basis for a social movement.

In response, a number of theologians have gravitated toward the metaphor of the techno–social–cultural cyborg as a better view of the posthuman self because of its emphasis on embodiedness and relationality, which are at the heart of the Christian tradition, where persons are enmeshed in relationships with others, with nature, and with God. Stephen Garner suggests that we are all "natural-born" cyborgs because we all form relationships with technology that expand the human mind outside of bodily limits. Technology, he writes, "is the environment in which we live and breathe and have our being."[8] He seeks a theology which includes the metaphor of cyborg because it not only accurately reflects reality, it also finds resonance with Christian theological understandings of the image of God, the Trinity, and the Incarnation (where there is "an inseparable connection between the self and other").[9] This emphasis on the communitarian dimension of theology provides hope for moving away from the transhumanist tendency toward individualism.[10] Similarly, Thweatt-Bates contends that the theological supports for the cyborg metaphor "provide a foundation for an ethics of relationship that is radically inclusive."[11] In her view, the metaphor of the cyborg accords well with the Christian theological emphasis on the social and historical embeddedness of the embodied human knower where there is communal accountability.[12]

It is important as we move in the direction of this technology and the tremendous implications and risks associated with it to clarify fully what we mean by responsibility. I have found that H.R. Niebuhr's concept of

[7] Ronald Cole-Turner, "Transhumanism and Christianity," in Cole-Turner, *Transhumanism and Transcendence*, 198.

[8] Garner, Stephen, "The Hopeful Cyborg," in Cole-Turner, *Transhumanism and Transcendence*, 89.

[9] Ibid., 92.

[10] Garner suggests that the *imago Dei* as a metaphor for the cyborg "contains an understanding of the interdependent, embodied relationships in which humanity is caught up in the natural world; a call to agency that does not dehumanize or marginalize others; of the imperative toward beneficent agency, while also recognizing potential for maleficence; and of a recognition that human activity, while questing for the transcendent, is still rooted in the finite." Ibid., 97.

[11] J. Jeanine Thweatt-Bates, "Artificial Wombs and Cyborg Births," in Cole-Turner, *Transhumanism and Transcendence*, 109.

[12] Thweatt-Bates, *Cyborg Selves*, 11.

the responsible self grounds the moral notion of responsibility quite well and is in keeping with the embodied, relational understanding of the post-human self which is articulated by those who employ the cyborg metaphor. However, Niebuhr goes further, eliminating any shadows, by emphasizing the significance of accountability.

For Niebuhr, the meaning of responsibility in moral agency involves four elements. First, all of our actions have the characteristic of being responses to action upon us. We are not simply self-active beings as seems to be the case in the transhumanist image. Rather, because we are social beings enmeshed in relations with others as the cyborg image attests, we are constantly confronted by action not subject to our control which demands response. This dynamic is part and parcel of what it means to live in society.

But such a response does not become a moral response, a response of a self, without the self's "interpretation" of the action upon it. This is a second element of what it means to be responsible. In order for any experience to become an experience of the self, the self must be actively and consciously involved. Uninterpreted responses bypass who we are as thinking, sentient, needful beings, that is, who we are as selves. Thus, a prior question for the responsible agent to the question "What shall I do?" is "What is going on?" The answer to the latter question will illuminate the interpretive framework the self will use as the basis for its answer to the former question.[13] And this framework is not simply conscious and rational, but incorporates "the deep memories that are buried within us, of feelings and intuitions that are only partly under our immediate control."[14]

Responsible actions not only attempt to relate to learned interpretive frameworks, they are also forward-looking. The responsible self anticipates reactions to its actions and incorporates them into its decision-making. This is what it means to be a "time-full" self. Accountability, the third element in responsibility, arises when the self takes those reactions into account. Niebuhr writes: "Responsibility lies in the agent who stays with his action, who accepts the consequences in the form of reactions and looks forward in a present deed to the continued interaction."[15]

[13] Niebuhr, *The Responsible Self*, 62–63.
[14] Ibid., 63.
[15] Ibid., 64.

This give and take points to the fourth element of the symbol of responsibility: social solidarity.[16] The community in which the self responds and is responded to is a continuing community of interaction.[17] The importance of this is that responsibility requires self-continuity, which in turn necessitates some continuity in the community in which the self lives. Here, the social self becomes the responsible self, particularly within the self's interpretive framework.

At this point, Niebuhr introduces the triadic pattern "self—social companion—cause," so important for his moral phenomenology. Niebuhr associates the emergence of cause with the self's need for meaning and worth.[18] Fundamentally, this is related to the dawning of consciousness or intelligence in the self in which the self tries to make sense out of its environment.[19] That which corresponds to the self's organic need for meaning and worth is a cause which becomes the value-center from which the self derives such meaning and worth.[20] In addition, Niebuhr remarks that causes, like values such as science, the nation, and religion which give the self multiple centers of value, may be legion.[21]

The significance of this triadic pattern, "self–social companion–cause," for the moral life is evident in Niebuhr's insistence that the self can consider itself responsible in the full sense only in this context: "Responsiveness now becomes responsibility in the sense of accountability when response is made not to one being alone but to that being as related with the self to a third reality."[22] The reason for this is twofold.

First, the cause enlarges our interpretive framework; it becomes the transcendent referent by which the self achieves independence from its social companions in that it can interpret the actions of others from a more universal vantage point than the one that is present in the immediate situation. Second, the cause broadens the scope of the self's responsibility. Niebuhr gives the example of Western societies' efforts to educate their children for responsible citizenship:

[16] Ibid., 65.

[17] You find this emphasis on social solidarity within a community of interaction among a number of transhumanists. For example, Margaret Wertheim argues that "If cyberspace teaches us anything it is that the worlds we conceive ... are communal projects requiring ongoing communal responsibility." (Wertheim, *The Pearly Gates of Cyberspace*, 304).

[18] Niebuhr, *Radical Monotheism and Western Culture*, 118.

[19] Niebuhr, *The Responsible Self*, 61.

[20] Niebuhr, *Radical Monotheism and Western Culture*, 110–113.

[21] Niebuhr, *Christ and Culture*, 38.

[22] Niebuhr, *The Responsible Self*, 62.

> We do not think that they will become responsible if they are related simply to their fellow citizens. They must also have direct connection with their country and its cause—what it stands for—so that they can interpret the actions of their fellow citizens in the context of the national intention; so that they will not be subject to the tyranny of the immediate instance and the present moment.[23]

The third element of the triad completes the context or "moral order" in which the self's interpretations and responses are always set. Niebuhr contends that by acting only in light of this triadic structure, the self "has become not only a responsive but also an accountable self."[24]

This notion of accountability as a key feature of responsibility is an important addition to the discussion of responsibility and removes the shadow associated with the other views of the self. Most of those who are affiliated with the transhumanist movement and those who affirm the "hopeful cyborg" both want the human and posthuman to be responsible, moral selves and act in ways that embody the fundamental values they espouse—to affirm the dignity of all. In my view, Niebuhr's view of the responsible self provides a stronger grounding for those hopes because it recognizes the embodied, relational self and affirms the accountability required of persons living in social solidarity with one another. The assumption of responsibility and accountability in a context of social solidarity is especially important in an environment where humanity and posthumanity must bear the costs and risks involved with changing or overcoming some aspects of human finitude, risks which transhumanists and their critics recognize as concerns.[25] Responsibility must anticipate the potential changes of technology and should consider the social and political implications of such changes. This latter concern points to the importance of the notion of political community which governs the context for responsible action.

A Shadow of Political Community

As noted earlier, there is a desire on the part of most transhumanists to develop technology responsibly and ethically. They recognize certain risks associated with human enhancement technologies and want to be sure

[23] Ibid., 85–86.

[24] Ibid., 65.

[25] For a transhumanist perspective on risks, see Nick Bostrom, "Existential Risks: Analyzing Human Extinction Scenarios and Related Hazards." For a critical perspective, see Francis Fukuyama, *Our Posthuman Future*.

that these are developed in accordance with some shared values. According to the Transhumanist Declaration:

> We need to carefully deliberate how best to reduce risks and expedite beneficial applications. We also need forums where people can constructively discuss what should be done, and a social order where responsible decisions can be implemented. ... Policy making ought to be guided by responsible and inclusive moral vision, taking seriously both opportunities and risks, respecting autonomy and individual rights, and showing solidarity with and concern for the interests and dignity of all people around the globe. We must also consider our moral responsibilities towards generations that will exist in the future.[26]

This desire for careful deliberation and the establishment of a social order where ethical decisions can be made is a laudable goal. The challenge is to determine which political social order would allow this to happen.

Some transhumanists are libertarian and want very little interference from government or religion with their pursuit of "morphological freedom." Such an individualistic framework is not shared by all. Implied in the earlier statement is some version of a liberal social contract view of political community, which Nick Bostrom says historically has resulted in an increase in those deemed as full moral persons with equal rights, including people without property, persons of color, and women. He sees no reason why the social contract could not be extended to include posthumans. "We can work to create more inclusive social structures that accord appropriate moral recognition and legal rights to all who need them, be they male or female, black or white, flesh or silicon."[27]

While it is true that the social contract has been more inclusive, the inclusion has not been universal—a crucial element of its imperfection. In the face of the levels of extreme inequality we have today, inequality that critics of transhumanism fear will get worse, it is not clear that the social contract can meet the challenge. A key reason for this is that contract views of political community, whether of libertarian or liberal democratic bent, are inherently individualistic. Persons exist as individuals prior to membership in society; in a contract society persons are autonomous agents who relate to one another to promote their mutual values, interests, or advantages. You can see this perspective throughout transhumanist

[26] "Transhumanist Declaration."

[27] Bostrom, "In Defense of Posthuman Dignity," 210.

views with their emphasis on the rights of individuals to use technology as they see fit, with a wide expanse of individual freedom. Bostrom and others do recognize that we may need to have some constraints for the sake of others, but his assumption is that posthumans will be more moral and more willing to place restraints upon themselves to stave off the catastrophic or existential risks that a free-wheeling pursuit of human enhancement technologies might unintentionally bring about.[28] Many critics, such as Ted Peters, are not so sure.[29]

James Hughes, who affirms the transhumanist vision, does challenge the libertarian views of some transhumanists and wants to complement the liberal social contract view of others with an increased democratization of society, which controls, regulates, and insures the equitable distribution of human enhancement technologies. Hughes argues that only new forms of transhuman citizenship and democracy will make us freer, more equal, and more united:

> To create a transhuman democracy we will have to establish a new definition of citizenship, a "cyborg citizenship," based on personhood rather than humanness. With cyborg citizenship we can deal with the scary boundary-crossers, the cyborgs, the animal-human hybrids, the genetically engineered kids, the clones and the robots. We can add some more chairs at the table.[30]

Leaving aside the theory of personhood that underlies Hughes' claim, the liberal democratic social contract theory still underlies this conception of political community. It is still individualistic in the sense that individuals band together in order to pursue their common interests. Its emphasis on equality and democracy does add a more communitarian layer to the libertarian vision but does not fully take into consideration the inherent social, interdependent nature of humanity (highlighted in the previous

[28] Bostrom ("Dignity and Enhancement," 203–204) writes: "By choosing to recognize these values and to treat the world accordingly, they would be accepting some constraints on their actions. It is by accepting such constraints that they could build, or rather *cultivate* their Plastic World into something that has greater value than a daydream. It is also by accepting such constraints—perhaps only by doing so—that it would be possible for them to preserve their own Dignity as a Quality. This dignity would not consist in resisting or defying the world. Rather, theirs would be a dignity of the strong, consisting in self-restraint and the positive nurturance of both internal and external values."

[29] Peters, Ted, "Progress and Provolution: Will Transhumanism Leave Sin Behind?" in Cole-Turner, Transhumanism and Transcendence, 82.

[30] Hughes, *Citizen Cyborg*, 79.

section) upon which to ground an ethic of responsibility and obligation.[31] Nor does his notion of cyborg citizenship adequately account for the role of the state in securing equity and fairness. Here I contend that the Christian tradition's concept of covenant provides a stronger foundation for political community than the social contract tradition because it offers the best possibility for securing the moral obligations of society and the state that Hughes and other transhumanists seek.

The Christian tradition of covenant contends that the primary covenant is the covenant between God and all of creation. All individuals are intended for membership in it whether they affirm it or not. Thus everyone potentially comes under its promise and protection. This inclusive covenant implies two things. In contrast to the self-sufficient individual in the contract view, a covenantal society emphasizes the essential social, relational, interdependence of life: individuals only find meaning or become selves—persons—in community. There is, thus, a strong connection between an ethics of responsibility and a covenantal conception of community.

On the other hand, this inclusive covenant bestows worth on each individual as God's creation and as members of the same moral community. As a result, members must recognize the worth of each member for his/her own sake and not for his/her value as an instrument to one's own self-interest, what Bostrom might refer to as dignity as recognition.[32] They must respect the intrinsic worth of each member equally. They must be faithful in the fulfillment of their responsibilities to each other. This is true of all individuals, human and posthuman.

It is true that covenant and contract have common roots. They are both agreements based on the mutual consent of the parties involved. They both emphasize individual freedom and mutual responsibility. But they are not identical. Covenant recognizes that the well-being of one is intricately (not just instrumentally) connected to the well-being of others. One cannot pursue one's own good without the concern for the well-being of others. Freedom is important in a covenantal framework, but freedom is

[31] Kathryn Hayles provides this critique: "The basis for a shared society—that is, the contract that reciprocally benefits both participants—breaks down when those who have nothing to give outnumber those who have much to give, for any contract must then be unequal and hence unfair to the privileged" (Hayles, *How We Became Posthuman*, 223).

[32] Bostrom, "Dignity and Enhancement."

conceived relationally rather than individualistically. Instead of meaning the capacity to choose among goods, true freedom means the ability to commit oneself to a more inclusive cause.

Of course, most transhumanists are secular and would balk at my use of religious language. However, the language of covenant has a long history in debates about political community internationally (e.g. United Nations Covenants) and certainly in American political life. Niebuhr suggests that the pattern of covenant or federal society reflected in Puritan thought was one of the guiding frameworks for the founders of our American democracy and constitution.[33] There is a strong sense of mutual obligation and responsibility inherent in this covenantal view of political community, a far cry from the limited obligations of the contract tradition that undergirds the transhumanist view, where having freedom means exercising power and authority over one's life without undue interference by others, especially government influenced by religious conservatives. Nick Bostrom writes: "A liberal democracy should normally permit incursions into morphological and reproductive freedoms only in cases where somebody is abusing these freedoms to harm another person."[34]

Political theorist Daniel J. Elazar contends that a covenantal view of political community is broader. Those bound by covenant "are obligated to respond to each other beyond the letter of the law rather than to limit their obligations to the narrowest contractual requirements."[35] In this view of political community, we are bound not just by geographical proximity but by pledges and promises we make to one another to work together for a common cause or end. In political community, we entrust ourselves to each other and risk our well-being to the power others have over us and we over them. The state, the organization with the highest political authority and power, not only provides protection for its citi-

[33] Niebuhr, "The Idea of Covenant in American Democracy." In general, one might question how the idea of covenant can engage a broad, religiously diverse audience, including individuals who do not subscribe to a higher power. With regard to those of other religious traditions, the Council for a Parliament of World Religions has endorsed *Toward a Global Ethic* (https://parliamentofreligions.org) whose principles embody most of the responsibilities inherent in the notion of a covenant political community. With regard to those who do not subscribe to a higher power, the U.N. International Covenant on Civil and Political Rights embodies explicitly covenantal language and the U.N. Declaration on Human Rights embodies implicitly most of the principles of a covenant political community. Thus, the idea of covenant and the responsibilities it entails is not a foreign concept to these groups.

[34] Bostrom, "In Defense of Posthuman Dignity," 210.

[35] Elazar, *Covenant and Polity in Biblical Israel*, 31.

zens but also becomes a repository of our common deliberation concerning the good of the community. It also serves as the agent for assuring that the wishes of the community are carried out, which even transhumanists seek.

The moral obligations of the state in a covenantal political community include, first of all, an obligation to be inclusive of all its citizens in its deliberations and actions.[36] In political community, we desire the government to act in ways that promote the well-being of all. This is why we invest them with authority and power over all the persons within their boundaries. Governments fail to live up to that trust when they rule only on behalf of some citizens and not all, especially the powerless and the poor. The democratic state has the obligation to insure that its rules and laws apply to all and to provide the means for all citizens and groups to have a meaningful voice in the political process.

Second, the covenant-oriented state is obligated to seek the common good. As implied earlier, the common good includes both the social conditions that allow all persons to live life more fully and the opportunity for all to participate and contribute to those social conditions. The common good is different from the sum total of individual goods implicit in the contract tradition, which assumes that everyone seeking his or her own good would result in the common good. Rather, the common good recognizes our interdependence; the good of each person is bound up with the good of the community. Members cannot pursue their own goods without thought for the good of others. Instead, they pursue these goods responsibly in ways that promote not only their own good but also the good of others in the community. It is common also in the sense that all members share in both its benefits and its burdens. If some individuals do not share in the benefits of social and technological advance, then our interdependence becomes a locus of domination and exploitation, exactly what critics of transhumanism fear. Thus, one of the social conditions necessary for the common good is justice, the proper distribution of society's benefits and burdens. Moreover, to achieve justice, the state may have to restrain or regulate certain activities, which even transhumanists acknowledge must occur, to guarantee that they contribute to the common good rather than undermine it.

[36] For a fuller treatment of the moral obligations of political community from a covenantal perspective, see Allen, *Love & Conflict*, chapter 9.

Finally, in addition to affirming the common good, covenant allows us to unify conceptions of justice as both virtue and law in a concept of political community based upon mutual trust and fidelity to a common cause. In some ways the unifying function of covenant resembles the unifying nature of responsibility as a paradigm for moral relationships. Responsibility brings together the teleological and deontological forms of ethics; it takes abstract principles of the good and the right and makes them concrete in human relationships. Covenant does the same with justice. Neither justice as a virtue nor justice as law is complete by itself. A covenantal view of justice stresses the internal disposition to act justly ("virtue") and the faithfulness to obey the moral law ("rules of justice"). It also provides the transcendent norm by which all systems of human justice are measured: the cause of true human (and posthuman) fulfillment for all intended in God's creating and covenanting activity.

The Church as Shadow

In the previous two sections, I argued that the concepts of the self and political community that inform many transhumanists and their desire to pursue human enhancement technologies responsibly and equitably were shadows, imperfect or faint representations, of the persons and communities in which we are enmeshed. Theological and ethical considerations about the responsible self and political community drawn from the Christian tradition can provide some corrective which could satisfy some of the criticisms Christian theologians and ethicists have of transhumanism. In this final section, I want to talk about the relationship between the transhumanist movement and the church. Continuing with the metaphor of shadow, I want to argue that the church must be a shadow, a constant companion and follower, in the debates about human enhancement technologies in order to bring its collective theological and ethical wisdom to the table regarding the appropriate development of such technology, and to shed light on any dark or dangerous shadows that might emerge.

Niebuhr argues that "The Church lives and defines itself in action vis-à-vis the world."[37] God has called the church community to increase the love of God and neighbor (broadly defined) in the world. Yet theologians and churches have understood in different ways the nature and purpose of the church in relation to the culture (or techno-culture for our purposes)

[37] Niebuhr, *The Purpose of the Church and Its Ministry*, 26.

in which it lives, according to how much tension they perceive in the relationship.[38] Some see the church as an alternative community, embodying distinctive Christian values and practices that differ markedly from those in the culture around it. Its relationship to the broader society is often one either of withdrawal from the culture or witness against it.[39] One could argue that those Christian theorists who are labeled pejoratively by some transhumanists as "bioconservatives" or "bioLuddites" embody this kind of approach to efforts toward morphological freedom.

For others, the church is a covenant community. Members of the church bind their lives together, entrust themselves to one another, and promise to care for each other. Therefore, they have mutual obligations and responsibilities to create a community in which all can thrive. But the church is not only in covenant with one another but also with the God who calls the community into being. God's covenant is inclusive and bestows worth on each person as God's creation and as members of the same moral community. As the Apostle Paul writes, "There is neither Jew nor Gentile, neither slave nor free, nor is there male and female, for you are all one in Christ Jesus."[40] (We might even add that there is also neither human nor posthuman.) Thus to be faithful to its calling, the church must be an inclusive community, welcoming all, showing all the same regard and care demonstrated by God.

This view of the church has implications for the mission of the church in the world: to proclaim God's inclusive covenant both in word and deed. It must model an inclusive community that embodies equity, justice, and compassion for all individuals, human and posthuman, providing for their spiritual, moral, and social needs. It should also practice the same compassion to those outside the church and advocate a vision of society that fully values all individuals.

In its relationship to transhumanism, the church can neither ignore nor dismiss outright human enhancement technologies, suggesting that all such technological developments are outside its purview or that they violate what it sees as essential to what it means to be human. Ted Peters is right when he writes, "No, we should not play God in the promethean sense. But we should play human in the *imago Dei* sense—that is, we should understand ourselves as created co-creators and press our scientific

[38] See Richard H. Niebuhr's classic text, *Christ and Culture.*

[39] See Stassen et al., *Authentic Transformation*, chapters 1, 4.

[40] Galatians 3:28.

and technological creativity into the service of neighbor love, of beneficence."[41] Thus, the church has the responsibility to shadow those developments to draw parallels to Christian understandings of the world and to highlight both the positive and negative potential such technologies have for the flourishing of creation.

In its document "Fearfully and Wonderfully Made: A Policy on Human Biotechnologies," the National Council of Churches recognizes this responsibility:

> Inasmuch as we are responsible for tending God's creation, scientific endeavor is proper for a Christian because one should know as much as one can about what one is responsible for.
>
> *We are responsible*. As member churches, we bear witness to what God intended the Church and the world to be. We share a sense of urgency that all will share the fruits of the new and emerging biotechnologies. In our stewardship of the creation, we lift high the concept of the common good—that we live in a covenant community with responsibility for one another.[42]

The church should be grateful to those theologians and ethicists who have invested time and energy into understanding the transhumanist visions of the self and society and sought to discern parallels with the Christian theological tradition, including beliefs about creation, the image of God, the Trinity, the Incarnation, and the eschaton.[43] They have taken the time to "shadow" transhumanist proponents of human enhancement technologies and to learn from them instead of automatically assuming that they are unworthy of such time and effort. Not only have they appreciated the positive potential such technologies have for human flourishing, they have also drawn attention to their potential risks and to weaknesses in the transhumanist visions.

To fulfill its mission the broader church must encourage its members to do the same. The church must address these developments head-on, demonstrating that it is thoughtful in its response to them. The church must take a proactive role here instead of just reacting. The church should take what Ted Peters calls a proleptic stance, which he defines as "taking creative and transformative action in the present stimulated by our vision

[41] Peters, *Playing God*, 197.

[42] "Fearfully and Wonderfully Made: A Policy on Human Biotechnologies."

[43] The edited volume by Cole-Turner, *Transhumanism and Transcendence*, is an excellent resource in this regard. See also Mercer and Trothen, *Religion and Transhumanism*.

of the future."[44] In some ways, the reactive response is understandable given the speed and complexity of scientific and technological developments. But by seeking to develop appropriate theological and ethical engagement with these developments the church won't be considered merely a nay-sayer. Rather, they will become partners in a common search for knowledge and the common good, creating "a world friendly to scientific, ecological, and humanistic values."[45]

Moreover, the shadow metaphor calls upon the church to live up to its prophetic role: to bring its mission of sharing the gospel and love of neighbor to the table to illuminate the developments in human enhancement technologies so that the potential dark sides do not emerge. Ted Peters warns that "It would be a grave injustice indeed if the shadow side of genetic science and medical technology created new forms of neglect, injustice, and marginalization."[46]

Conclusion

While a focus on the future of the human is in order and part of the church's mission, the church's efforts to embody God's inclusive covenant must include challenging those social conditions and structures that leave so many in want of the basic necessities for human flourishing. The Church must always be asking the question as to how this will affect the entire community, not just the privileged few. In an era of limited funding, we need to be sure that the transhumanist desires of the scientific and technological communities do not usurp all other ethical and communal considerations and responsibilities. This is the fear of many: somehow the connection with finance capitalism and its focus on profits will exacerbate the inequality that already exists by pouring resources into technological investments in our posthuman future, rather than resolving the problems that keep so many of the human population from realizing their full potential.[47] The church must be an advocate for the disenfranchised and

[44] Peters, *Playing God*, 210.

[45] Grassie, William, "Millennialism at the Singularity: Reflections on the Limits of Ray Kurzweil's Exponential Logic," in Hansell and Grassie, *H+/-: Transhumanism and Its Critics*, 266.

[46] Peters, *Playing God*, 212.

[47] Langdon Winner argues that we need to think about what it means to be human, and rather than focus on "human nature" we need to focus on the "human condition" (Winner, "Resistance is Futile: The Posthuman Condition and Its Advocates," in Baillie and Casey, *Is*

challenge scientific and technological communities, many of whom are members of Christian congregations, to focus their energies on resolving these problems as well.

Thus the church must continue its prophetic role to draw attention to those places where the posthuman visions of transhumanism falter and to shed light on those places where there is commonality with the Christian vision of the human and posthuman future. As Ted Peters suggests, that future may require a radical act of divine grace to occur fully,[48] but it does not mean that the efforts on the part of the church, God's "created co-creators," do not play some significant role. In many ways, technology reflects—or is the shadow of—the cultural/religious values of society. Transhumanists draw upon particular humanistic values present in society, such as individual freedom and liberal democratic rights, by which they hope to shape future technological developments. The church can also help to shape any shadows these technologies cast by emphasizing the fuller meaning of these values in dialogue with the scientific, technological, and political communities in which the church and its members live. These values include the social nature of free, responsible, and accountable selves intricately enmeshed in relationship with others, a conception of the common good where all individuals can flourish, and a more transcendent understanding of justice as both a virtue and norm for all individuals and institutions. We need to shadow the developments to be sure that the more pessimistic and negative potential of the shadow does not gain the upper hand, and does not leave an attenuated or vestigial remnant of the human in its wake.

Bibliography

Allen, Joseph. 1984. *Love & Conflict: A Covenantal Model of Christian Ethics*. Nashville: Abingdon Press.

Baillie, Harold W., and Timothy K. Casey, eds. 2004. *Is Human Nature Obsolete? Genetics, Bioengineering, and the Future of the Human Condition*. Cambridge, MA: MIT Press.

Human Nature Obsolete?, 405). See also Ted Peters who suggests that the connection between transhumanist goals and laissez-faire capitalism and its emphasis on profits is a major concern. (Peters, "Progress and Provolution: Will Transhumanism Leave Sin Behind?" in Cole-Turner, *Transhumanism and Transcendence*, 72).

[48] Peters, Ted, "Transhumanism and the Posthuman Future," in Hansell and Grassie, *H+/-: Transhumanism and Its Critics*, 162.

Bostrom, Nick. 2002. Existential Risks: Analyzing Human Extinction Scenarios and Related Hazards. *Journal of Evolution and Technology* 9 (1): 1–36.
———. 2005a. Transhumanist Values. *Journal of Philosophical Research* 30 (Supplement): 3–14.
———. 2005b. In Defense of Posthuman Dignity. *Bioethics* 19 (3): 202–214.
———. 2008. Dignity and Enhancement. In *Human Dignity and Bioethics: Essays Commissioned by the President's Council on Bioethics*,ed. Edmund D. Pellegrino, Adam Schulman, and Thomas W. Merrill, 173–206. Washington, DC. https://repository.library.georgetown.edu/bitstream/handle/10822/559351/human_dignity_and_bioethics.pdf?sequence=1&isAllowed=y
Cole-Turner, Ronald, ed. 2011. *Transhumanism and Transcendence: Christian Hope in an Age of Technological Enhancement*. Washington, DC: Georgetown University Press.
———. 2015. Going Beyond the Human: Christians and Other Transhumanists. *Dialog: A Journal of Theology* 54 (1): 20–26.
Elazar, Daniel. 1995. *Covenant and Polity in Biblical Israel: Biblical Foundations and Jewish Expressions*. New Brunswick: Transaction Publishers.
"Fearfully and Wonderfully Made: A Policy on Human Biotechnologies." *National Council of Churches*. November 8, 2006. http://nationalcouncilofchurches.us/common-witness/2006/biotech.php
Fukuyama, Francis. 2002. *Our Posthuman Future: Consequences of the Biotechnology Revolution*. New York: Farrar, Strauss and Giroux.
Hansell, Gregory R., and William Grassie, eds. 2011. *H+/-: Transhumanism and Its Critics*. Philadelphia: Metanexus Institute.
Hayles, Katherine. 1999. *How We Became Posthuman: Virtual Bodies in Cybernetics, Literature, and Informatics*. Chicago: University of Chicago Press.
Hughes, James. 2004. *Citizen Cyborg: Why Democratic Societies Must Respond to the Redesigned Human of the Future*. New York: Basic Books.
Kurzweil, Ray. 2006. *The Singularity Is Near: When Humans Transcend Biology*. New York: Penguin Books.
Mercer, Calvin, and Tracy J. Trothen, eds. 2015. *Religion and Transhumanism: The Unknown Future of Human Enhancement*. Santa Barbara: Praeger.
Niebuhr, H. Richard. 1951. *Christ and Culture*. New York: Harper & Row.
———. 1954. The Idea of Covenant in American Democracy. *Church History* 23 (2): 126–135.
———. 1956. *The Purpose of the Church and Its Ministry*. New York: Harper & Row.
———. 1960. *Radical Monotheism and Western Culture*. New York: Harper & Row.
———. 1963. *The Responsible Self: An Essay in Christian Moral Philosophy*. New York: Harper & Row.

Peters, Ted. 2003. *Playing God: Genetic Determinisms and Human Freedom.* 2nd ed. New York: Routledge.

Stassen, Glen H., D.M. Yeager, and John Howard Yoder. 1996. *Authentic Transformation: A New Vision of Christ and Culture.* Nashville: Abingdon Press.

Thweatt-Bates, Jeanine. 2012. *Cyborg Selves: A Theological Anthropology of the Posthuman.* Farnham: Ashgate Publishing.

"Transhumanist Declaration." *Humanity+.* March, 2009. http://humanityplus.org/philosophy/transhumanist-declaration/

Wertheim, Margaret. 2000. *The Pearly Gates of Cyberspace.* New York: W.W. Norton.

Young, Simon. 2006. *Designer Evolution: A Transhumanist Manifesto.* Amherst: Prometheus Books.

CHAPTER 13

Rivalry, Control, and Transhumanist Desire

Ysabel Johnston

Introduction

This generation is in a unique position; questions about technological advancement, once only abstract, now take on new meaning. Old science fiction scenarios are now real possibilities to be grappled with—would we want to expand our life spans indefinitely, upload our brains to computers, or vastly improve our intelligence and physical abilities? However, these possibilities are often framed as looming realities rather than inquiries. There seems to be an implicit assumption that human beings desire certain things, or at least *should* desire certain things. This assumption misses a fundamental question: *why* do we desire, or don't desire, these things?

The transhumanism movement treats the desire to overcome biological limitations not only as acceptable, but normal, dignified, and even moral. From their perspective, it is obvious that people want to extend their lives indefinitely and vastly improve their capabilities. They seem self-evidently desirable. Any objections to the transhumanists' desires are framed as

Y. Johnston (✉)
Albert Gnaegi Center for Health Care Ethics, Saint Louis University,
St Louis, MO, USA
e-mail: ysabel.johnston@slu.edu

S. Donaldson, R. Cole-Turner (eds.), *Christian Perspectives on Transhumanism and the Church*, Palgrave Studies in the Future of Humanity and its Successors,
https://doi.org/10.1007/978-3-319-90323-1_13

resulting from fear, delusion, or ignorance of technological advance. In other words, you *should* want what they want; if you don't there's something wrong with you.

However, the work of René Girard offers a compelling response to the desirability of transhumanist projects. He argues that desires result from relational states—they are social in nature, not objective or universal as some transhumanists seem to suggest. Given this theory, how should transhumanist goals and desires be understood? This chapter will consider the transhumanist desires for progress and control of biology from a theological, Girardian perspective. Such an outlook gives us a guideline to evaluate human desires: they are good insofar as they result from a healthy relationship with God. In contrast, desires for progress and control can be symptomatic of rivalry with God.

First, this chapter will discuss the basic goals, desires, and views of human nature characterizing transhumanism. Though it's difficult to characterize an entire movement, some salient themes can be drawn out. Transhumanist ideology is paradoxical in its view of human nature: humanity is viewed as a formless work in progress, but also as fundamentally oriented toward desiring specific goods (namely, the goods of control and progress). Next, basic tenets of Girard's thought will be explicated. Humans are imitative by nature, and desires result from this kind of social imitation. Jean Michel Oughourlian has a helpful Girardian account of Genesis demonstrating the process whereby imitated desire leads to rivalry, specifically rivalry between humans and God. Furthermore, the transhumanist view of human nature is proto-Nietzschean in assuming power or progress is fundamentally desirable. A Girardian analysis situates this consuming desire for power not as fundamental, but as a result of rivalry.

This analysis is best understood as suggesting a kind of virtue ethic—its applications will focus on the attitudes of those pursuing technology rather than the rightness or wrongness of specific technologies. Rivalistic relationships with God lead to a mad pursuit of control, while a right relationship with God allows one to yield control. Only within this right relationship can we wisely evaluate and pursue new possible uses of technology. Any act to advance the human species should originate from a healthy relationship with the Creator of the human species.

Context: Transhumanism

The goal of transhumanism is to extend human potential in a variety of ways.[1] According to Nick Bostrom, "Humanity+ formally defines it based on Max More's original definition as follows: The intellectual and cultural movement that affirms the possibility and desirability of fundamentally improving the human condition through applied reason, especially by developing and making widely available technologies to eliminate aging and greatly enhance human intellectual, physical, and psychological capacities."[2]

Transhumanists situate their roots in secular humanist thinking,[3] but they radicalize this thinking. They seek to grab the reigns of biological evolution and determine humanity's evolutionary process. This is acceptable because "transhumanists view human nature as a work-in-progress, a half-baked beginning that we can learn to remold in desirable ways."[4] There is no underlying, consistent human essence that delineates what we can and can't change. On the transhumanist account, human nature has always been a work in progress; therefore, there is no issue with humans determining their own progression.

They hope improvements go so far as to allow for the existence of a new type of human: the posthuman. The posthuman would have benefitted and adapted by means of the new technological and medical breakthroughs; the changes could be so drastic that the posthuman bears little resemblance to current humans.[5] Ray Kurzweil predicts an event when technology advances so rapidly that human life as we know it will be radically and irreversibly altered.[6] This point, deemed "the singularity," will be followed by the saturation of the universe with intelligence.[7] As others have noted, this posthuman phase works as a kind of transhumanist eschatology.[8]

[1] Cf. Bostrom, "Transhumanist FAQ 3.0."

[2] Ibid. The second part of the definition describes transhumanism as "The study of the ramifications, promises, and potential dangers of technologies that will enable us to overcome fundamental human limitations, and the related study of the ethical matters involved in developing and using such technologies."

[3] Bostrom "Transhumanist Values," 4.

[4] Ibid.

[5] Ibid.

[6] Kurzweil, *The Singularity is Near*, 2–5.

[7] Ibid., 21, 364–366.

[8] Bishop, "Transhumanism, Metaphysics, and the Posthuman God," 707.

Much transhumanist literature is concerned with clarifying what they foresee the transhuman or posthuman future to be.[9] Some assume objections will evaporate once this future is clarified, as everyone will see the intrinsic desirability of this posthuman future. Those who *still* don't desire this future after gaining a clear view of it must, according to them, be plagued by fear. When Aubrey de Grey discusses his goal of ending aging, he notes that "people who are totally rational and open to discourse on any other matter approach the topic of defeating aging with a resistance to debate that virtually defies description."[10] He deems that these people are entrenched in a "pro-aging trance,"[11] a kind of psychological coping mechanism which views aging and death as natural, and perhaps good. In de Grey's view, the only way to make sense of such strong opposition to his project is to pathologize it. This illustrates the transhumanists' general view that resistance to their projects is due to irrationality.

Transhumanists are therefore making strong claims about what humans want, and they're committed to the fact that the technologies and advancements they propose can fulfill these desires. What are these desires? Simply put, they desire more control over biology in order to seek progress. These two goals are intimately linked, and tend to collapse into one. Humans should have more control, that is, more freedom and ability to progress. The content of "progress" tends to be allowing more exercise of individual freedom. They desire to control in order to progress, and progression entails more control.

Though transhumanists are not a homogeneous group, it's fair to say that the movement is characterized by the desirability of control and progress. They take these to be basic human desires because of the view that human nature is essentially driven to improve itself and shape itself. Therefore, transhumanism is operating on a very specific definition of human nature and human desire. On one hand, they see human nature as elastic—able to be shaped and changed into the posthuman. On the other hand, human nature is static—it always wants control and progress. These are considered immutable.

[9] Kurzweil, *The Singularity Is Near*, 21; Bostrom "Transhumanist FAQ 3.0."
[10] De Grey and Rae, *Ending Aging*, 11.
[11] Ibid., 9–11.

Girard: Desire Is Socially Based

René Girard is a Franco-American thinker whose work has been far-reaching, impacting literary criticism, sociology, and theology. Theologians such as Raymund Schwager[12] and James Alison[13] have realized the importance of his theories for Christology and soteriology. His concepts of mimetic desire, violence, and mimetic rivalry, will serve as important tools to evaluate transhumanism.

For Girard, the only "static" aspect of human nature is the fact that we imitate. We borrow from those to whom we are exposed, particularly those closest to us.[14] Language, expressions, mannerisms, and especially desires are mimetic—our human disposition is to imitate each other. Girard focuses primarily on the mimetic nature of desire, asserting that desire is not determined by ourselves; "we do not each have our own desire, one really our own."[15] Desires do not arise individually—they are dependent on and arise from community.

On Girard's account, desire can be distinguished from physical needs for food, water, and so on. He writes that "biologically determined appetites and needs, which are common to both men and animals … stand in contrast to *desire* and *passion*, which are exclusively human."[16] Desire is something beyond mere need for food and shelter—it is a passion that comes about through imitation. Someone can want water, but this want doesn't count as the kind of imitated passion that Girard describes.

Desire operates in a triangular fashion, involving a subject, a model, and an object of desire.[17] The subject is a person who comes to imitate the model, thereby adopting the model's desire—or at least what the subject *perceives* to be the model's desires. The model displays desire for the object, which could be a material object such as a painting, keepsake, or a piece of clothing. The object of desire could also be abstract, such as power, recognition, or knowledge. It's not the case that the subject desires this object, whatever it may be, because the object is intrinsically desirable. Rather, the subject desires the objected precisely because the model desires

[12] See Schwager, *Must There Be Scapegoats?*
[13] See Alison, *The Joy of Being Wrong*.
[14] Girard, *I See Satan Fall Like Lightning*, 15.
[15] Ibid.
[16] Girard, *The One by Whom Scandal Comes*, 4.
[17] Girard, *I See Satan Fall Like Lightning*, 9.

it. “We assume that desire is objective or subjective, but in reality it rests on a third party who gives value to the objects.”[18] Desire is due to the subject imitating the model.

Girard’s main points are thus the following: humans are imitative by nature, and desires are imitated and therefore are not independent of social relation. Desire arises by having role models to imitate. Girard makes this clear when he writes that “truly to desire, we must have recourse to people about us; we have to borrow their desires.”[19] His last point is that mimesis can easily lead to violent rivalries—the discussion of this point will be picked up later.

Why take Girard’s thought seriously? First, the concept of humans as imitative beings has long held sway in philosophy. Aristotle also claimed that “Imitation is natural to man from childhood, one of his advantages over the lower animals being this, that he is the most imitative creature in the world, and learns at first by imitation.”[20] Girard expands upon this basic insight, offering a more thoroughgoing account of the process and effects of imitation.

Furthermore, there is a growing body of empirical research on imitation and its effects on desire that offers support for Girard’s claims.[21] Infant research suggests babies learn through imitation, and are naturally social animals inclined to imitate.[22] Some have suggested that mirror neurons are the neural mechanisms by which imitation occurs.[23] Though there isn’t the space within this chapter for a more lengthy discussion of the scientific, empirical, and cultural research undergirding Girard’s thought, it would be a mistake to assume Girard’s claims lack scientific merit.

Still, many take issue with the view that *all* desire is due to imitation—it seems to lend itself to a relativistic view where there is no objective value, only value that is constructed socially. There are two relevant ways one could respond to this concern: the first response departs somewhat from Girard and maintains that there are a few fundamental, “objective” desires. Other desires could be imitated and socially based as Girard claims, but not *all* desire is due to imitation. The second response embraces Girard’s claim that *all* desire is due to imitation, but offers a kind of theological ground that reveals the falsity of the distinction

[18] Ibid.
[19] Ibid., 15.
[20] Aristotle, “Poetics” (1448b5) in Aristotle, *The Basic Works of Aristotle*, 1457.
[21] See Garrels, *Mimesis and Science*.
[22] Ibid. See Part 1, Chapter 3.
[23] Ibid. See Part 1, Chapter 2.

between socially constructed desire and objective desire. Because a relationship with God is the original relationship constituting humanity, desires can be weighed against each other on the basis of coherence with the desires of God. Desire is imitated, but doesn't fall into relativism because it originates with God. The following arguments will be based on this perspective. However, one need not view *all* desire as mimetic in order to problematize the transhumanists' claim that the desires for control and progress are fundamental.

The "Genesis" of Desire

An obvious question arises with Girard's account of human desire—where do you get "original" desire? If desire is imitated, what was the first imitation? How did the first model's desire arise, if not imitated? Either an endless cycle of imitation extends back through time or there was a starting point.

Girard scholar Jean-Michel Oughourlian traces this starting point back to Genesis.[24] God's own desires were humanity's first desires. Adam and Eve originally imitated God the Father in Eden—this healthy imitation occurred in a spirit of obedience and shamelessness.[25] In fact, Oughourlian argues that the term "desire" is somewhat misleading in this context, as that seems to imply a psychological fixation. Rather, these imitated desires were more like shared interests—for example, Adam and Eve imitated God's care for creation.

This healthy imitative relationship shifted when Eve encountered the snake. It generates desire for the fruit from the tree of the knowledge of good and evil by convincing Eve that God desires its fruit only for Himself.[26] The snake claims that "God knows that when you eat from it your eyes will be opened, and you will be like God, knowing good and evil" (Genesis 3:5). In this instance, God seems to find the fruit desirable, therefore Eve begins to desire it. This case illustrates the triangular nature of desire: God is the model, Eve is the subject, and the fruit is the object. The snake frames God's prohibition against eating the fruit not as concern, but as covetousness.[27] The snake introduces a false understanding of God—a selfish, coveting God.

[24] Alison also offers a Girardian reinterpretation of original sin, atonement, and the resurrection in *The Joy of Being Wrong*.

[25] Oughourlian, *The Genesis of Desire*, 52.

[26] Ibid., 62.

[27] Ibid.

It's important to note here that it is not *merely* the fruit that is desired. The locus of the desire is really what the fruit will bring about—the acquisition of the divine knowledge of good and evil. This knowledge is associated with *being like God* because it represents an attribute of God—His omniscience.[28] Thus, Eve's imitation of God becomes what Girard deems *metaphysical desire*. This is a desire to overtake the *being* of the model, not merely to possess an object they desire.[29] No longer does Eve want what God wants; she wants to *be God*, in a sense. Oughourlian's description of such a desire is powerful: "[O]ur feelings take on an extreme intensity; we want to melt into the other, take his place, rob him of his very being, of the secret of his luminous aura, of the autonomy that we dream of that seems to be his."[30] As he notes, metaphysical desire involves more intense desire, which in this case amounts to a kind of psychological fixation.

Adam and Eve's initial relationship with God did not involve metaphysical desire. The snake entered the story and changed Eve's perception of God with a lie. As Oughourlian notes, "this lie consists in making her believe that God desires that particular tree because it confers on Him divine knowledge and power and that it is because He desires it that he wants to hold it back from Adam and Eve—rather than protect them from death."[31] This false understanding makes the object of desire God's *being*. God becomes both the model and the object of desire, amounting to metaphysical desire. This effectively puts Eve, then Adam, in a rivalistic relationship with God. No longer is their imitation of God peaceful—it becomes competitive.

Girard explicates how the mimetic nature of desire generally raises the possibility for rivalry and violence, though there are important distinctions between human–human rivalries and human–divine rivalries.

Through imitation, humans start adopting the same objects of desire. If the object is scarce, those who imitate each other (and therefore desire the same thing) can easily enter metaphysical desire.[32] In this case, the desire is intensified and the imitators become rivals vying for the same object. The desire can heighten to a point where the rivals seek to overtake the other's being. Therefore, these rivals are competitors after the same

[28] Ibid., 48.
[29] Girard and Williams, *The Girard Reader*, 40.
[30] Oughourlian, *The Genesis of Desire*, 23.
[31] Ibid., 65.
[32] Ibid., 14.

object, and also after each other's being. The rivalry begins and perpetuates through imitation, not through each person independently becoming interested in the same object. Because imitation is humanity's natural disposition, Girard considers these types of rivalries unavoidable.[33]

In the case of Adam and Eve's rivalry with God, the rivalry is unidirectional. They set God as a rival, but God does not do the same to them. They buy the snake's lie and *perceive* God as covetous of His own knowledge and power. Since these are aspects of God's character, the mimesis shifted and produced *metaphysical desire.* Adam and Eve came to desire God's very being. This is a scarce object of desire—it's impossible for both God *and* Adam and Eve to obtain God's being. This conflict made a right relationship with God impossible, so Adam and Eve were cast out of the garden. Gil Bailie, another Girard scholar and theologian, summarizes the case aptly: "The fall, then, involves two things: mimetic desire for the fruit and mimetic rivalry and resentment toward the divine."[34]

Transhumanism in Light of Mimetic Theory

Ramez Naam, a technologist, award-winning author, and transhumanist, writes: "'Playing God' is actually the highest expression of human nature. The urges to improve ourselves, to master our environment, and to set our children on the best path possible have been the fundamental driving forces of all of human history."[35] According to this view, the essential orientation of humans is to want and seek progress. Desire to control biology is normal, fundamental, and commendable, on Naam's account. How would Girard or Oughourlian respond to this?

The theoretical frame I've provided offers two responses to Naam. First, it is problematic to view humanity as fundamentally oriented toward the kind of progress and control characteristic of transhumanism. Mimetic theory destabilizes any kind of claim that humans hold fundamental desires independent of social context. The notion that desire comes about through a subject imitating a model problematizes the claim that "playing God" is a natural expression of human nature. Rather, this orientation could have come about through mimesis. As described in the previous

[33] Ibid.

[34] Bailie, *Violence Unveiled*, 137–140 (opening sections of chapter 7, "A Text in Travail").

[35] Naam, *More than Human*. Ray Kurzweil also cites this quote in agreement in *The Singularity is Near*.

section, Oughourlian offers a possible account for the origin of the orientation toward control and progress. This kind of orientation has a source and a history, and certainly isn't a healthy orientation.

This emphasis on the desire for control as fundamental is reminiscent of a Nietzschean view of human nature, as noted by Jeffrey Bishop.[36] Nietzsche views humanity as fundamentally oriented toward power—both gaining power and deploying power.[37] This "will-to-power" is itself unexplained, but is used to explain all other actions and phenomena. Even actions that seem kind, humble, and even selfless can be explained as expressions of power.[38] This power is not pejorative for Nietzsche, in fact, expression of the will-to-power can be life-affirming. He often refers to creative, consuming practices as healthy expressions of the will-to-power. Those who embrace power are described as having "a powerful physicality, a blossoming, rich, even overflowing health, together with that which is required for its preservation: war, adventure, the hunt, dance, athletic contests, and in general everything which includes strong, free, cheerful-hearted activity."[39] We can hear echoes of this mentality in the transhumanists' hopes for the posthuman future.[40]

Similarly, transhumanists see humanity as fundamentally oriented toward progress. Their view of progress correlates with the Nietzschean view of power in that progress effectively means further ability to express power. Control of biology is a deployment of power on biology, and thus allows for the development and use of new posthuman powers and abilities. The cause of this orientation toward progress and control cannot and need not be explained further, according to transhumanists, just as the cause of the will-to-power can't be explained further. Bishop notes, "the ontology of thinkers like Harris and Bostrom is a power ontology, where power circulates in the stops and starts of evolutionary biology."[41] In this picture, power naturally circulates through evolution, so using human power to direct evolution is merely a continuation of a natural orientation of reality.

Girard directed significant attention toward a critique of Nietzsche's views of the will-to-power and violence.[42] Unlike Nietzsche, his thought is

[36] Bishop, "Transhumanism, Metaphysics, and the Posthuman God," 705–707.

[37] Nietzsche, *On the Genealogy of Morality*, 11.

[38] Ibid., 15. His critique of Christianity claims that the exultation of virtues such as humility is itself an act of power.

[39] Ibid., 16.

[40] Bostrom "Transhumanist FAQ 3.0."

[41] Bishop, "Transhumanism, Metaphysics, and the Posthuman God," 706.

[42] See Girard, *I See Satan Fall Like Lightning*, Chapter 14.

able to account for *why* the will-to-power, or perhaps we should say "will-to-progress," seems pervasive, but is not. The desire for power or progress is not fundamental, but is imitated. A mimetic understanding of human nature brings the transhumanist paradoxical view of human nature into relief. Calling progress and control "fundamental driving forces" suggests they are basic, unchangeable desires that all humans experience. On Girard's account, desires are relationally based and historically situated, not unchanging aspects of human nature. Therefore, we can't claim any desire as "natural" and justify it on that basis. Instead, we can notice the relationships and context that give rise to the desire. Because desires are socially based, we can ask what *kind* of relationship produces the desire for progress and control.

This leads to the second point: transhumanists seem to be operating on a false view of God. They commit the same error as Eve in characterizing God primarily as a powerful, controlling being.

If by "playing God" Naam means overcoming biological limitations and mastering our environment, then God is an overcomer and a master. God is unlikely to yield control, and God is unlikely to allow progress. God seems to be tyrannical in nature, fearful that humans will try to take control of what has been Divinely ordained and deemed "natural." Control and progress seem scarce and valuable if so carefully protected by this Divine Being. With this view of God, it's easy to want to dethrone the tyrant and allow for our own freedom. The rivalry drives us to want to beat God and control our own morality and mortality.

This competitive attitude declares that *we* should be in control, that we could shape nature better. Accordingly, Charles Rubin notes the transhumanists' view of themselves as "consummate problem solvers who have come to understand how much better things would have been if someone had asked us how they should be arranged."[43] Their belief that humans can design and create much better than God is highly competitive—humans can plan better, design better, and control better. God is viewed as intending creative powers only for Himself.

Even more problematic is the Nietzschean and transhumanist view that *dignity* is situated in the creative, controlling force. Rubin argues this is implicit in transhumanist ideology: "Indeed, it is precisely this rejection of resignation, this capacity for perpetual problem-solving and self-overcoming, that makes human beings worthy of respect in the first place,

[43] Rubin, "Human Dignity and the Future of Man," 157.

that gives us our *dignity*."[44] Human dignity is not bestowed by God; dignity is found in competing with God. On this view, Adam and Eve's eating of the fruit is the most dignified act in Genesis.

One could object that many transhumanists are also atheists, and therefore could not be operating on the views of God I've mentioned. There are two responses to this: first, it's not necessary to consciously assent to God's existence in order to be in a rivalry with Him. It's not as if humans suddenly lose all relationality with God simply because they don't believe He exists. Second, the theology animating transhumanist projects is implicit. They would not declare themselves as having a Nietzschean ideology, and likely wouldn't explicitly describe God as controlling and competitive. Rather, these notions lurk behind transhumanist thought and projects. For example, the desire to free humanity from mortality and perfect it into posthumanity implies that the eschaton will not come to fruition unless humans create it. Their eschatology involves *becoming* this posthuman god who can handle omnipotence better than the deistic god who tried to do so, but set us on a "half-baked beginning."[45]

Transhumanists are trying to best the deistic, controlling, omnipotent god who made the mistake of allowing frailty and suffering. Stanley Hauerwas notes in *God, Medicine, and Suffering* that "although this seems contradictory, in fact the view of an all-powerful but basically deistic God fits nicely with the understanding of the necessity of humankind's taking control of its destiny."[46] In other words, the view of God as primarily powerful and removed encourages humans to imitate this kind of character, and thus remove themselves from God and compete against the Divine. These concepts "fit nicely" when we understand humankind's rivalry with our own Creator.

In sum, transhumanism is problematic insofar as it arises from a rivalistic relationship with God. The view of god implicit in transhumanist ideology is a Nietzschean god of power that has attempted to hold humanity back by imposing biological limitation. Rivalry with this god generates desires to control our existence by overcoming the biological. Therefore, we can become the ultimate self-creators. These desires lead us away from the true character of God and from a healthy relationship with God.

[44] Ibid., 157.
[45] Bostrom "Transhumanist Values," 4.
[46] Hauerwas, *God, Medicine and Suffering*, 48.

Christ as Imitative Exemplar

This analysis of transhumanists as imitators of a false god raises some confusion. If imitating god is problematic and leads to rivalry, it seems we must stop encouraging imitation of Christ. Perhaps we should rename Thomas á Kempis' *Imitation of Christ* as *Rivalry with Christ*. However, rejecting imitation of Christ doesn't acknowledge the possibility of healthy imitation. Contrary to this concern, imitation of Christ is exactly what will prevent the perpetuation of rivalry, violence, and competition with God.

Christ himself is both God and imitating God. He is able to reveal the character of God and is at the same time an embodied example of a right relationship between God and humanity. Because Christ *is* God, he has no misconceptions of the nature of God's desires. He does not view God as competitive or controlling. He does not overemphasize God's power above other characteristics.

Rivalry is characterized by the urge to *be* God while healthy imitation is characterized by the urge to *be like* God, drawing near to God by striving to have a similar, loving character. The first wants God to be replaced by the self. The second wants the self to be in conjunction with God. The first relies upon misconceptions of God's desires. The second has a more accurate conception of God's desires. A rivalry is not merely an inclination to be *like* God, but a competition that must result in one winner—either God or oneself. It involves metaphysical desire. In this sense, it is distinct from healthy imitations of God as described by Thomas á Kempis and other theologians.

Girard distinguishes between mimetic desire characteristic of Jesus and mimetic desire characteristic of Satan: "The difference is that Satan imitates God in a spirit of rivalry. Jesus imitates God in a spirit of childlike and innocent obedience and this is what he advises us to do as well."[47] Christ both demonstrates healthy imitation of God, and clarifies God's character for us.[48] He effectively denounces the false view of God as controlling and competitive. He makes it clear that God does not view humanity as rivals, but simply desires a right relationship with humanity. God's power is not a separate characteristic, wholly distinct from God's love. Through imitating Christ and setting him as our moral exemplar our very desires are shaped. It's not only the case that we gain practical wisdom and knowledge—our desires are set in accordance with God's.

[47] Girard and Williams, *The Girard Reader*, 197. Excerpt from Girard's essay "Satan."
[48] Girard, *I See Satan Fall Like Lightning*, 123.

However, it's somewhat misleading to assert that Christians can consciously choose Christ as an exemplar, and then follow through with imitation of Christ and adoption of God's desires. It's not as simple as that. On Girard's account, imitation operates primarily in the subconscious and cannot merely be decided upon. One can make the decision to imitate Christ, setting him as the exemplar, yet fail to actually proceed in subconscious mimesis of him. This is because mimesis requires continued exposure to the model, or exemplar, and is an ongoing process. The decision to imitate Christ seems to be necessary, but not sufficient for the inculcation of the desires of God.

Therefore, the Church is absolutely necessary to foster and encourage mimesis of Christ. The Church constantly gathers imitators of Christ, who therefore imitate one another and reinforce Christ's values and desires in their own hearts. The continual reorientation of the self toward God in worship, the reminders of Christ's narrative, the sacramental encounters with God, all these operate on the subconscious in such a way as to transform desire. In his introduction to *Desiring the Kingdom*, James K.A. Smith cleverly describes a shopping mall in such a way as to highlight its similarity to a religious space of worship.[49] His contention is similar to Girard's—these communities and spaces of worship shape our desires. The Church must offer a counter to cultural institutions and communities that lead away from the desires of God, and must instead model God's desires for one another. Desire can heighten dangerously due to rivalries, but it can also grow in the continual imitation of Christ in the community of the Church body.

Conclusion

While Girard's account of desire and rivalry doesn't give us the resources to determine whether specific technologies are morally wrong, it does supply us the resources to evaluate the desire for control that underlies transhumanism. However, we must be careful not to totalize this diagnosis. Though rivalries with God surely foster desires for control and progress, we are in no position to definitively claim that all transhumanists are in rivalries with God. Transhumanists could have differing motives and desires. Furthermore, it takes a profound well of wisdom to grasp an individual's motivations and reasons for desiring a particular thing. We, as the Church, shouldn't pretend to espouse that type of wisdom for people we don't know well.

[49] Smith, *Desiring the Kingdom*, 19–22.

At the same time, the kind of eschaton implicit in transhumanism—a posthuman, disburdened existence where contingency is stamped out and human liberty reigns over biology—is not the end Christ points us to. Christ was not out for his own liberty. Christ was fully open to the contingency of the messy world he inhabited. The desire for such a disburdened existence does not derive from imitation of Christ, but from competition with a god who withholds goods.

It would be unwise to ignore the impact of such rivalries. We should use this framework as a tool for understanding likely issues in the overall transhumanist movement. It can be offered as a method of self-evaluation to Christian transhumanists pursuing technological advancement. We can use it to evaluate our own desires for control and advancement. We can use it carefully to sharpen each other within close relationships. We should ask deeper questions about what is meant by "progress" and encourage each other to evaluate our own motivations. We should be wary when it seems someone desires control more than closeness with his or her Creator.

Bibliography

Alison, James. 1998. *The Joy of Being Wrong: Original Sin Through Easter Eyes.* New York: Crossroad.

Aristotle. 1941. *The Basic Works of Aristotle*, ed. Richard McKeon. New York: Random House.

Bailie, Gil. 1995. *Violence Unveiled: Humanity at the Crossroads.* New York: Crossroad.

Bishop, Jeffrey P. 2010. Transhumanism, Metaphysics, and the Posthuman God. *Journal of Medicine & Philosophy* 35 (6): 700–720.

Bostrom, Nick. 2005. Transhumanist Values. *Journal of Philosophical Research* 30 (Supplement): 3–14.

———. Transhumanist FAQ 3.0. *Humanity+*. http://humanityplus.org/philosophy/transhumanist-faq/. Accessed 20 Aug 2015.

de Grey, Aubrey, and Michael Rae. 2007. *Ending Aging: The Rejuvenation Breakthroughs That Could Reverse Human Aging in Our Lifetime.* New York: St. Martin's Press.

Garrels, Scott R. 2011. *Mimesis and Science: Empirical Research on Imitation and the Mimetic Theory of Culture and Religion.* East Lansing: Michigan State University Press.

Girard, René. 2001. *I See Satan Fall Like Lightning.* Maryknoll: Orbis Books.

———. 2014. *The One by Whom Scandal Comes.* East Lansing: Michigan State University Press.

Girard, René, and James G. Williams. 1996. *The Girard Reader*. New York: Crossroad.

Hauerwas, Stanley. 1990. *God, Medicine, and Suffering*. Grand Rapids: Eerdmans.

Kurzweil, Ray. 2005. *The Singularity Is Near: When Humans Transcend Biology*. New York: Viking.

Naam, Ramez. 2005. *More than Human: Embracing the Promise of Biological Enhancement*. New York: Broadway Books.

Nietzsche, Friedrich Wilhelm. 1998. *On the Genealogy of Morality*. Translated with Introduction and Notes by Maudemarie Clark and Alan J. Swensen. Indianapolis: Hackett Pub.

Oughourlian, Jean-Michel. 2010. *The Genesis of Desire*. East Lansing: Michigan State University Press. Electronic.

Rubin, Charles. 2008. Human Dignity and the Future of Man. In *Human Dignity and Bioethics: Essays Commissioned by the President's Council on Bioethics*, ed. Edmund D. Pellegrino, Adam Schulman, and Thomas W. Merrill, 155–172. Washington, DC: The President's Council on Bioethics. https://repository.library.georgetown.edu/bitstream/handle/10822/559351/human_dignity_and_bioethics.pdf?sequence=1&isAllowed=y

Schwager, Raymund. 1987. *Must There Be Scapegoats? Violence and Redemption in the Bible*. San Francisco: Harper & Row.

Smith, James K.A. 2009. *Desiring the Kingdom: Worship, Worldview, and Cultural Formation*. Grand Rapids: Baker Academic.

CHAPTER 14

Epilogue: The Church—Bigger Bangs Are Coming

Steve Donaldson

My son, if you accept my words and store up my commands within you, turning your ear to wisdom and applying your heart to understanding—indeed, if you call out for insight and cry aloud for understanding, and if you look for it as for silver and search for it as for hidden treasure, then you will understand the fear of the Lord and find the knowledge of God. For the Lord gives wisdom; from his mouth come knowledge and understanding.
Proverbs 2:1–6

So, who does what? When it comes to wisdom, for example, is the emphasis on human seeking or divine dispensation? Are they inseparable and, if so, what does that mean? Christians—from the very beginning—have struggled to determine the roles they are expected to play in the divine–human narrative. While acknowledging the enabling power of God, we find reason to believe in our own autonomy—not only because we are

S. Donaldson (✉)
Samford University, Birmingham, AL, USA
e-mail: sfdonald@samford.edu

S. Donaldson, R. Cole-Turner (eds.), *Christian Perspectives on Transhumanism and the Church*, Palgrave Studies in the Future of Humanity and its Successors,
https://doi.org/10.1007/978-3-319-90323-1_14

encouraged to make God a choice but because our entire experience suggests that all of our decisions influence both who and what we are and are to be. This tension is palpable through the refrains of prophetic and apocalyptic literature that fill the pages of the Bible, calling for a change of heart and action on the one hand but professing on the other that, without divine intervention, no meaningful transformation is possible.

But this is much more than a theological exercise. For the religious, evolution has been a theological exercise; Big Bang cosmology was a theological exercise. Yet the ability to expand physical and cognitive boundaries is tautological in a world of ever-increasing scientific understanding and technological expertise, and it seems clear that many of those limit-removing activities will be deemed beneficial for us and our descendants. Here we move from the merely theoretical into the realm of change where we will actually live. And such radical changes will also almost certainly affect conceptions of what it means to be human, influence views of spirituality and deity, impact the church, and modify ways of living and interacting.

For Christians, the questions are these: Can a climate be created in which churches engage a transhumanist future positively or must the church resist? Is resistance futile? To what extent will the church's response be based on presumptions about the respective roles of God and people? Can Christians actually lead the way toward common and valuable perspectives with respect to forthcoming change? Can our insights be communicated to a broader audience?

While we are as yet unable to answer those questions, it seems safe to say that it cannot merely be "business as usual" if the church is to play a meaningful part in this drama. This doesn't mean the church abandons its salvific ministry but casts it in light of a historic and eschatological framework that transcends what has been an all too common reactive, rather than proactive, stance on major cultural issues. The concerns here are complex and far-reaching and will present major problems for individuals accustomed to casting everything in simple binary terms.

This book, then, is a call for the church to begin the attempt to clarify its thinking regarding its place in a transhuman world. Obviously that can only occur if it is, indeed, thinking. Unfortunately, much of the rhetoric to date has a knee-jerk and sometimes muddled character. Why, for instance, would Christians not embrace a transhumanist paradise which promises immortality, no suffering, no sadness, improved cognition, enhanced physical abilities, and perfect health when those are the very things they claim to be looking forward to in heaven? Is it because they have failed to note a certain logical inconsistency or simply because they have leaned toward the apocalyptic vision?

Ultimately, such questions can only be addressed in light of core Christian belief and practice. Yet while Christians have more or less agreed on just what it is that constitutes that core, there has never been unanimity. In fact, in that realization lies the power to move forward with the contentious issues of transhumanism. Balancing prophetic and apocalyptic views requires looking ahead, but neither too far (thus forgetting where we are now and what God might have in mind for us in the relatively near future) nor not far enough (thereby ignoring a greater destiny). Although we might harbor the sentiment that we are just where God intends for us to be, we must also acknowledge that Jesus' injunction to "Be perfect, therefore, as your heavenly Father is perfect" (Matthew 5:48) implies that we are never exactly where we are meant to be.

CHAPTER 15

Epilogue: People of Vision, Communities of Discernment

Ron Cole-Turner

Everyone I know accepts the idea that technology is going to continue to advance. We even think we have a pretty good idea *how* it will advance, maybe not in the distant future, but at least in the next few years. By knowing where we are today, we can foresee at least in broad terms where key technological developments are most likely to take us.

Consider robotics, nanotechnology, information technology, biotechnology, or medicine. The more we know about where things stand today, the more confident we are that we know what is coming. Robots will expand from industrial applications until they become companions and domestic aides. Computers will interact directly with brains. The human genome will become editable and babies designable. And so it goes.

It is one thing to foresee the coming of new technologies. It is another thing entirely to predict their full human impact. How will these technologies change our social world? How will they change our bodies? How will they change our brains, our thoughts, our feelings, and our sense of who

R. Cole-Turner (✉)
Pittsburgh Theological Seminary, Pittsburgh, PA, USA
e-mail: coleturn@pts.edu

S. Donaldson, R. Cole-Turner (eds.), *Christian Perspectives on Transhumanism and the Church*, Palgrave Studies in the Future of Humanity and its Successors,
https://doi.org/10.1007/978-3-319-90323-1_15

we are as individuals and what we are collectively as human beings? Predicting the next technology is relatively easy compared to predicting how humans will react to it and be transformed by it.

Get people together in groups, like churches, and the problem of prediction becomes even more daunting. And yet we have to ask: What kind of church will we have in the future? This is not a question about technology. It is not even a question about how technology will change people in general. We are asking how it will change the Christian faith and the experience of the Christian life. How will it change the meaning of being a follower of Jesus Christ? How will it change what we expect from our encounter with the transformative power of the Gospel?

No one can claim today to know how to answer these questions. Will Christians rise to address the technological challenges of our times? Will pastors show that they are aware of the technological forces reshaping our lives? Will congregations take on the task of equipping the saints for *Christian living today*? If so, then spiritual insight and technologically savvy discernment will come bit by bit, conversation by conversation, sermon by sermon, as actual Christians create concrete answers by the way they live their lives in communities of faith.

In every age, the church at its best has seen its task, not as telling people what to think or do, but as equipping them for a life defined by active spiritual and moral discernment. And the best churches are pretty good at nurturing people to be spiritually alive to the world, ready to distinguish between its authentic joys and God-given blessings, on the one hand, and the vices, compulsions, degradations, and worldly anxieties that keep us from flourishing as God's people, on the other. We are to be as wise as serpents and as innocent as doves; in the world but not of it, as Jesus put it.

Now the world wants to get into your body and brain. Its technology, pre-loaded with all the values of a secular, market-driven age, wants to enter your bloodstream, modulate your brain, mess with your mood, and make you capable of doing new things. In response, today's Christians need a new kind of spiritual discernment. What technology offers can be enticing. Some of it is truly positive and life-affirming. But who among us does not worry about it, wondering about its hidden costs in depriving us of some good, old-fashioned practices or some real in-the-flesh facetime?

How can a Christian make sense of our moment in time, with its insane pace of newness and its endless updates? How can a Christian be a good parent, watching our children's faces glow in the light of omnipresent screens and wondering what they are getting and what they are missing? How can we find ourselves being made new by God when we are constantly being made different by technology?

The solution for Christians is not to withdraw into tech-free zones (even if they existed). For us, technology is neither satanic nor messianic. It will not destroy us or save us. It is simply part of our world today, that part that is redefining all the other parts. It is not some alien, demonic, or intergalactic invader. It is us. It is our inventions reinventing us. It has always been part of the human story. The peculiar fate of our generation, however, is to live at a time when technological novelty seems to change us so fast that we become a blur even to ourselves.

By itself, the idea that we are being changed should not frighten Christians. We are a people defined by our expectation of change. We hope for it. But for that very reason, technology introduces a special concern for Christians. How is God changing us? How is technology changing us? And how are the two related? This is exactly where discernment enters in. Technology is not all good or all bad. For the Christian, what makes technological enhancement good or bad is whether it fits within the more profoundly important spiritual and moral transformation that God is effecting in our lives. Our challenge is to discern what fits.

There is no book, no theology text, and no creed that spells out what fits and what does not. Where do we turn in the Bible to find out if taking pills to enhance cognition is morally different from drinking coffee? What verses guide us in making decisions about when brain/chip implants are acceptable?

What God gives us is each other. A vibrant, contemporary church is a community of moral and spiritual discernment, a safe place where people can share what God is doing in their lives. How are we being made new? How are we being set free from our compulsions and fears? How is the God of abundant grace working in all things for good in our lives?

We come together to share the meaning of letting ourselves be transformed until we become new creations. We share how technology helps and hurts us in that transformation. We are honest with each other, revealing our uncertainties and our struggles, asking for the guidance of the community and the prayers of friends. Step by step, we discern how we are to live as Christians when technology is redefining everything, including us. And in all things, we seek to grow together in one body, that of our living Savior. This is the church that is needed now more than ever.

Index[1]

[1] Note: Page numbers followed by 'n' refer to notes.

S. Donaldson, R. Cole-Turner (eds.), *Christian Perspectives on Transhumanism and the Church*, Palgrave Studies in the Future of Humanity and its Successors,
https://doi.org/10.1007/978-3-319-90323-1

C

Q

R

The manufacturer's authorised representative in the EU is Springer Nature Customer Service Centre GmbH, Europaplatz 3, 69115 Heidelberg, Germany. If you have any concerns regarding our products, please contact ProductSafety@springernature.com

Printed and bound by CPI Group (UK) Ltd, Croydon, CR0 4YY
30/09/2026
02225195-0004